Lecture Notes in Mathematics　　　1764

Editors:
J.-M. Morel, Cachan
F. Takens, Groningen
B. Teissier, Paris

Ana Cannas da Silva

Lectures on
Symplectic Geometry

Corrected 2nd printing 2008

 Springer

Ana Cannas da Silva
Departamento de Matemática
Instituto Superior Técnico
1049-001 Lisboa
Portugal
acannas@math.ist.utl.pt

and

Department of Mathematics
Princeton University
Princeton, NJ 08544-1000
USA
acannas@math.princeton.edu

ISBN: 978-3-540-42195-5   e-ISBN: 978-3-540-45330-7
DOI: 10.1007/978-3-540-45330-7

Lecture Notes in Mathematics ISSN print edition: 0075-8434
                              ISSN electronic edition: 1617-9692

Library of Congress Control Number: 2008924358

Mathematics Subject Classification (2000): 53Dxx, 53D05, 53D20, 53-xx, 53-01

Corrected 2nd printing 2008
ⓒ 2001 Springer-Verlag Berlin Heidelberg

Cover design: WMXDesign GmbH, Heidelberg

Printed on acid-free paper

9 8 7 6 5 4 3 2 1

springer.com

Foreword

These notes approximately transcribe a 15-week course on symplectic geometry I taught at UC Berkeley in the Fall of 1997.

The course at Berkeley was greatly inspired in content and style by Victor Guillemin, whose masterly teaching of beautiful courses on topics related to symplectic geometry at MIT, I was lucky enough to experience as a graduate student. I am very thankful to him!

That course also borrowed from the 1997 Park City summer courses on symplectic geometry and topology, and from many talks and discussions of the symplectic geometry group at MIT. Among the regular participants in the MIT informal symplectic seminar 93–96, I would like to acknowledge the contributions of Allen Knutson, Chris Woodward, David Metzler, Eckhard Meinrenken, Elisa Prato, Eugene Lerman, Jonathan Weitsman, Lisa Jeffrey, Reyer Sjamaar, Shaun Martin, Stephanie Singer, Sue Tolman and, last but not least, Yael Karshon.

Thanks to everyone sitting in Math 242 in the Fall of 1997 for all the comments they made, and especially to those who wrote notes on the basis of which I was better able to reconstruct what went on: Alexandru Scorpan, Ben Davis, David Martinez, Don Barkauskas, Ezra Miller, Henrique Bursztyn, John-Peter Lund, Laura De Marco, Olga Radko, Peter Přibík, Pieter Collins, Sarah Packman, Stephen Bigelow, Susan Harrington, Tolga Etgü and Yi Ma.

I am indebted to Chris Tuffley, Megumi Harada and Saul Schleimer who read the first draft of these notes and spotted many mistakes, and to Fernando Louro, Grisha Mikhalkin and, particularly, João Baptista who suggested several improvements and careful corrections. Of course I am fully responsible for the remaining errors and imprecisions.

The interest of Alan Weinstein, Allen Knutson, Chris Woodward, Eugene Lerman, Jiang-Hua Lu, Kai Cieliebak, Rahul Pandharipande, Viktor Ginzburg and Yael Karshon was crucial at the last stages of the preparation of this manuscript. I am grateful to them, and to Michèle Audin for her inspiring texts and lectures.

Finally, many thanks to Faye Yeager and Debbie Craig who typed pages of messy notes into neat LaTeX, to João Palhoto Matos for his technical support, and to Catriona Byrne, Ina Lindemann, Ingrid März and the rest of the Springer-Verlag mathematics editorial team for their expert advice.

Berkeley, November 1998   *Ana Cannas da Silva*
and Lisbon, September 2000

Contents

Part XI Symplectic Toric Manifolds

Introduction

The goal of these notes is to provide a fast introduction to symplectic geometry.

A symplectic form is a closed nondegenerate 2-form. A symplectic manifold is a manifold equipped with a symplectic form. Symplectic geometry is the geometry of symplectic manifolds. Symplectic manifolds are necessarily even-dimensional and orientable, since nondegeneracy says that the top exterior power of a symplectic form is a volume form. The closedness condition is a natural differential equation, which forces all symplectic manifolds to being locally indistinguishable. (These assertions will be explained in Lecture 1 and Homework 2.)

The list of questions on symplectic forms begins with those of existence and uniqueness on a given manifold. For specific symplectic manifolds, one would like to understand the geometry and the topology of special submanifolds, the dynamics of certain vector fields or systems of differential equations, the symmetries and extra structure, etc.

Two centuries ago, symplectic geometry provided a language for classical mechanics. Through its recent huge development, it conquered an independent and rich territory, as a central branch of differential geometry and topology. To mention just a few key landmarks, one may say that symplectic geometry began to take its modern shape with the formulation of the Arnold conjectures in the 60's and with the foundational work of Weinstein in the 70's. A paper of Gromov [49] in the 80's gave the subject a whole new set of tools: pseudo-holomorphic curves. Gromov also first showed that important results from complex Kähler geometry remain true in the more general symplectic category, and this direction was continued rather dramatically in the 90's in the work of Donaldson on the topology of symplectic manifolds and their symplectic submanifolds, and in the work of Taubes in the context of the Seiberg-Witten invariants. Symplectic geometry is significantly stimulated by important interactions with global analysis, mathematical physics, low-dimensional topology, dynamical systems, algebraic geometry, integrable systems, microlocal analysis, partial differential equations, representation theory, quantization, equivariant cohomology, geometric combinatorics, etc.

As a curiosity, note that two centuries ago the name *symplectic geometry* did not exist. If you consult a major English dictionary, you are likely to find that *symplectic* is the name for a bone in a fish's head. However, as clarified in [105], the word *symplectic* in mathematics was coined by Weyl [110, p.165] who substituted the Latin root in *complex* by the corresponding Greek root, in order to label the symplectic group. Weyl thus avoided that this group connote the complex numbers, and also spared us from much confusion that would have arisen, had the name remained the former one in honor of Abel: *abelian linear group*.

This text is essentially the set of notes of a 15-week course on symplectic geometry with 2 hour-and-a-half lectures per week. The course targeted second-year graduate students in mathematics, though the audience was more diverse, including advanced undergraduates, post-docs and graduate students from other departments. The present text should hence still be appropriate for a second-year graduate course or for an independent study project.

There are scattered short exercises throughout the text. At the end of most lectures, some longer guided problems, called homework, were designed to complement the exposition or extend the reader's understanding.

Geometry of manifolds was the basic prerequisite for the original course, so the same holds now for the notes. In particular, some familiarity with de Rham theory and classical Lie groups is expected.

As for conventions: unless otherwise indicated, all vector spaces are real and finite-dimensional, all maps are smooth (i.e., C^∞) and all manifolds are smooth, Hausdorff and second countable.

Here is a brief summary of the contents of this book. Parts I–III explain classical topics, including cotangent bundles, symplectomorphisms, lagrangian submanifolds and local forms. Parts IV–VI concentrate on important related areas, such as contact geometry and Kähler geometry. Classical hamiltonian theory enters in Parts VII–VIII, starting the second half of this book, which is devoted to a selection of themes from hamiltonian dynamical systems and symmetry. Parts IX–XI discuss the moment map whose preponderance has been growing steadily for the past twenty years.

There are by now excellent references on symplectic geometry, a subset of which is in the bibliography. However, the most efficient introduction to a subject is often a short elementary treatment, and these notes attempt to serve that purpose. The author hopes that these notes provide a taste of areas of current research, and will prepare the reader to explore recent papers and extensive books in symplectic geometry, where the pace is much faster.

Part I
Symplectic Manifolds

A symplectic form is a 2-form satisfying an algebraic condition – nondegeneracy – and an analytical condition – closedness. In Lectures 1 and 2 we define symplectic forms, describe some of their basic properties, introduce the first examples, namely even-dimensional euclidean spaces and cotangent bundles.

Chapter 1
Symplectic Forms

1.1 Skew-Symmetric Bilinear Maps

Let V be an m-dimensional vector space over $\mathbb{R}$, and let $\Omega : V \times V \to \mathbb{R}$ be a bilinear map. The map Ω is **skew-symmetric** if $\Omega(u,v) = -\Omega(v,u)$, for all $u, v \in V$.

Theorem 1.1. (Standard Form for Skew-symmetric Bilinear Maps)
  Let Ω be a skew-symmetric bilinear map on V. Then there is a basis $u_1, \ldots, u_k, e_1, \ldots, e_n, f_1, \ldots, f_n$ *of V such that*

$$\begin{aligned} &\Omega(u_i, v) = 0\,, &&\text{for all } i \text{ and all } v \in V, \\ &\Omega(e_i, e_j) = 0 = \Omega(f_i, f_j)\,, &&\text{for all } i, j, \text{ and} \\ &\Omega(e_i, f_j) = \delta_{ij}\,, &&\text{for all } i, j. \end{aligned}$$

Remarks.

1. The basis in Theorem 1.1 is not unique, though it is traditionally also called a "canonical" basis.
2. In matrix notation with respect to such basis, we have

$$\Omega(u,v) = [\,{-}u{-}\,] \begin{bmatrix} 0 & 0 & 0 \\ 0 & 0 & \mathrm{Id} \\ 0 & -\mathrm{Id} & 0 \end{bmatrix} \begin{bmatrix} | \\ v \\ | \end{bmatrix}.$$

Proof. This induction proof is a skew-symmetric version of the Gram-Schmidt process.

Let $U := \{u \in V \mid \Omega(u,v) = 0 \text{ for all } v \in V\}$. Choose a basis $u_1, \ldots, u_k$ of U, and choose a complementary space W to U in V,

$$V = U \oplus W\,.$$

Take any nonzero $e_1 \in W$. Then there is $f_1 \in W$ such that $\Omega(e_1, f_1) \neq 0$. Assume that $\Omega(e_1, f_1) = 1$. Let

$$W_1 = \text{span of } e_1, f_1$$
$$W_1^{\Omega} = \{w \in W \mid \Omega(w, v) = 0 \text{ for all } v \in W_1\} \ .$$

Claim. $W_1 \cap W_1^{\Omega} = \{0\}$.

Suppose that $v = ae_1 + bf_1 \in W_1 \cap W_1^{\Omega}$.

$$\left. \begin{array}{l} 0 = \Omega(v, e_1) = -b \\ 0 = \Omega(v, f_1) = \ \ \ a \end{array} \right\} \quad \implies \quad v = 0 \ .$$

Claim. $W = W_1 \oplus W_1^{\Omega}$.

Suppose that $v \in W$ has $\Omega(v, e_1) = c$ and $\Omega(v, f_1) = d$. Then

$$v = \underbrace{(-cf_1 + de_1)}_{\in W_1} + \underbrace{(v + cf_1 - de_1)}_{\in W_1^{\Omega}} \ .$$

Go on: let $e_2 \in W_1^{\Omega}$, $e_2 \neq 0$. There is $f_2 \in W_1^{\Omega}$ such that $\Omega(e_2, f_2) \neq 0$. Assume that $\Omega(e_2, f_2) = 1$. Let $W_2 = \text{span of } e_2, f_2$. Etc.

This process eventually stops because $\dim V < \infty$. We hence obtain

$$V = U \oplus W_1 \oplus W_2 \oplus \cdots \oplus W_n$$

where all summands are orthogonal with respect to Ω, and where W_i has basis e_i, f_i with $\Omega(e_i, f_i) = 1$. $\square$

The dimension of the subspace $U = \{u \in V \mid \Omega(u, v) = 0, \text{ for all } v \in V\}$ does not depend on the choice of basis.

$\implies \quad k := \dim U$ is an invariant of (V, Ω) .

Since $k + 2n = m = \dim V$,

$\implies \quad n$ is an invariant of (V, Ω); $2n$ is called the **rank** of Ω.

1.2 Symplectic Vector Spaces

Let V be an m-dimensional vector space over $\mathbb{R}$, and let $\Omega : V \times V \to \mathbb{R}$ be a bilinear map.

Definition 1.2. The map $\tilde{\Omega} : V \to V^*$ is the linear map defined by $\tilde{\Omega}(v)(u) = \Omega(v, u)$.

The kernel of $\tilde{\Omega}$ is the subspace U above.

Definition 1.3. A skew-symmetric bilinear map Ω is **symplectic** (or **nondegenerate**) if $\widetilde{\Omega}$ is bijective, i.e., $U = \{0\}$. The map Ω is then called a **linear symplectic structure** on V, and (V, Ω) is called a **symplectic vector space**.

The following are immediate properties of a linear symplectic structure Ω:

- **Duality**: the map $\widetilde{\Omega} : V \xrightarrow{\sim} V^*$ is a bijection.
- By the standard form theorem, $k = \dim U = 0$, so $\dim V = 2n$ is **even**.
- By Theorem 1.1, a symplectic vector space (V, Ω) has a basis $e_1, \ldots, e_n, f_1, \ldots, f_n$ satisfying

$$\Omega(e_i, f_j) = \delta_{ij} \quad \text{and} \quad \Omega(e_i, e_j) = 0 = \Omega(f_i, f_j) .$$

Such a basis is called a **symplectic basis** of (V, Ω). We have

$$\Omega(u, v) = [-u-] \begin{bmatrix} 0 & \mathrm{Id} \\ -\mathrm{Id} & 0 \end{bmatrix} \begin{bmatrix} | \\ v \\ | \end{bmatrix},$$

where the symbol $\begin{bmatrix} | \\ v \\ | \end{bmatrix}$ represents the column of coordinates of the vector v with respect to a symplectic basis $e_1, \ldots, e_n, f_1, \ldots, f_n$ whereas $[-v-]$ represents its transpose line.

Not all subspaces W of a symplectic vector space (V, Ω) look the same:

- A subspace W is called **symplectic** if $\Omega|_W$ is nondegenerate. For instance, the span of e_1, f_1 is symplectic.
- A subspace W is called **isotropic** if $\Omega|_W \equiv 0$. For instance, the span of e_1, e_2 is isotropic.

Homework 1 describes subspaces W of (V, Ω) in terms of the relation between W and W^Ω.

The **prototype of a symplectic vector space** is $(\mathbb{R}^{2n}, \Omega_0)$ with Ω_0 such that the basis

$$
\begin{aligned}
e_1 &= (1, 0, \ldots, 0), & \ldots, \quad e_n &= (0, \ldots, 0, \overset{n}{1}, 0, \ldots, 0), \\
f_1 &= (0, \ldots, 0, \underset{n+1}{1}, 0, \ldots, 0), & \ldots, \quad f_n &= (0, \ldots, 0, 1) ,
\end{aligned}
$$

is a symplectic basis. The map Ω_0 on other vectors is determined by its values on a basis and bilinearity.

Definition 1.4. A **symplectomorphism** φ between symplectic vector spaces (V, Ω) and (V', Ω') is a linear isomorphism $\varphi : V \xrightarrow{\sim} V'$ such that $\varphi^* \Omega' = \Omega$. (By definition, $(\varphi^* \Omega')(u, v) = \Omega'(\varphi(u), \varphi(v))$.) If a symplectomorphism exists, (V, Ω) and (V', Ω') are said to be **symplectomorphic**.

The relation of being symplectomorphic is clearly an equivalence relation in the set of all even-dimensional vector spaces. Furthermore, by Theorem 1.1, every $2n$-dimensional symplectic vector space (V,Ω) is symplectomorphic to the prototype $(\mathbb{R}^{2n},\Omega_0)$; a choice of a symplectic basis for (V,Ω) yields a symplectomorphism to $(\mathbb{R}^{2n},\Omega_0)$. Hence, nonnegative even integers classify equivalence classes for the relation of being symplectomorphic.

1.3 Symplectic Manifolds

Let ω be a de Rham 2-form on a manifold M, that is, for each $p \in M$, the map $\omega_p : T_pM \times T_pM \to \mathbb{R}$ is skew-symmetric bilinear on the tangent space to M at p, and ω_p varies smoothly in p. We say that ω is closed if it satisfies the differential equation $d\omega = 0$, where d is the de Rham differential (i.e., exterior derivative).

Definition 1.5. The 2-form ω is *symplectic* if ω is closed and ω_p is symplectic for all $p \in M$.

If ω is symplectic, then $\dim T_pM = \dim M$ must be even.

Definition 1.6. A *symplectic manifold* is a pair (M,ω) where M is a manifold and ω is a symplectic form.

Example. Let $M = \mathbb{R}^{2n}$ with linear coordinates $x_1,\ldots,x_n,y_1,\ldots,y_n$. The form

$$\omega_0 = \sum_{i=1}^n dx_i \wedge dy_i$$

is symplectic as can be easily checked, and the set

$$\left\{ \left(\frac{\partial}{\partial x_1}\right)_p, \ldots, \left(\frac{\partial}{\partial x_n}\right)_p, \left(\frac{\partial}{\partial y_1}\right)_p, \ldots, \left(\frac{\partial}{\partial y_n}\right)_p \right\}$$

is a symplectic basis of T_pM.   ◇

Example. Let $M = \mathbb{C}^n$ with linear coordinates $z_1,\ldots,z_n$. The form

$$\omega_0 = \frac{i}{2} \sum_{k=1}^n dz_k \wedge d\bar{z}_k$$

is symplectic. In fact, this form equals that of the previous example under the identification $\mathbb{C}^n \simeq \mathbb{R}^{2n}$, $z_k = x_k + iy_k$.   ◇

Example. Let $M = S^2$ regarded as the set of unit vectors in $\mathbb{R}^3$. Tangent vectors to S^2 at p may then be identified with vectors orthogonal to p. The standard symplectic form on S^2 is induced by the inner and exterior products:

$$\omega_p(u,v) := \langle p, u \times v \rangle, \qquad \text{for } u,v \in T_p S^2 = \{p\}^\perp.$$

This form is closed because it is of top degree; it is nondegenerate because $\langle p, u \times v \rangle \neq 0$ when $u \neq 0$ and we take, for instance, $v = u \times p$. $\diamondsuit$

1.4 Symplectomorphisms

Definition 1.7. Let (M_1, ω_1) and (M_2, ω_2) be $2n$-dimensional symplectic manifolds, and let $\varphi : M_1 \to M_2$ be a diffeomorphism. Then φ is a ***symplectomorphism*** if $\varphi^* \omega_2 = \omega_1$.[1]

We would like to classify symplectic manifolds up to symplectomorphism. The Darboux theorem (proved in Lecture 8 and stated below) takes care of this classification locally: the dimension is the only local invariant of symplectic manifolds up to symplectomorphisms. Just as any n-dimensional manifold looks locally like $\mathbb{R}^n$, any $2n$-dimensional *symplectic* manifold looks locally like $(\mathbb{R}^{2n}, \omega_0)$. More precisely, any symplectic manifold (M^{2n}, ω) is locally symplectomorphic to $(\mathbb{R}^{2n}, \omega_0)$.

Theorem 8.1 (Darboux) *Let (M, ω) be a $2n$-dimensional symplectic manifold, and let p be any point in M.*
Then there is a coordinate chart $(\mathcal{U}, x_1, \ldots, x_n, y_1, \ldots, y_n)$ centered at p such that on $\mathcal{U}$

$$\omega = \sum_{i=1}^n dx_i \wedge dy_i.$$

A chart $(\mathcal{U}, x_1, \ldots, x_n, y_1, \ldots, y_n)$ as in Theorem 8.1 is called a **Darboux chart**.

By Theorem 8.1, the **prototype of a local piece of a $2n$-dimensional symplectic manifold** is $M = \mathbb{R}^{2n}$, with linear coordinates $(x_1, \ldots, x_n, y_1, \ldots, y_n)$, and with symplectic form

$$\omega_0 = \sum_{i=1}^n dx_i \wedge dy_i.$$

---

[1] Recall that, by definition of pullback, at tangent vectors $u, v \in T_p M_1$, we have $(\varphi^* \omega_2)_p(u,v) = (\omega_2)_{\varphi(p)}(d\varphi_p(u), d\varphi_p(v))$.

Homework 1: Symplectic Linear Algebra

Given a linear subspace Y of a symplectic vector space (V, Ω), its **symplectic orthogonal** Y^{Ω} is the linear subspace defined by

$$Y^{\Omega} := \{v \in V \mid \Omega(v, u) = 0 \text{ for all } u \in Y\} \ .$$

1. Show that $\dim Y + \dim Y^{\Omega} = \dim V$.

    Hint: What is the kernel and image of the map

    $$V \longrightarrow Y^* = \operatorname{Hom}(Y, \mathbb{R}) \ ?$$
    $$v \longmapsto \Omega(v, \cdot)|_Y$$

2. Show that $(Y^{\Omega})^{\Omega} = Y$.
3. Show that, if Y and W are subspaces, then

    $$Y \subseteq W \iff W^{\Omega} \subseteq Y^{\Omega} \ .$$

4. Show that:
    Y is **symplectic** (i.e., $\Omega|_{Y \times Y}$ is nondegenerate) $\iff Y \cap Y^{\Omega} = \{0\} \iff V = Y \oplus Y^{\Omega}$.
5. We call Y **isotropic** when $Y \subseteq Y^{\Omega}$ (i.e., $\Omega|_{Y \times Y} \equiv 0$).
    Show that, if Y is isotropic, then $\dim Y \leq \frac{1}{2} \dim V$.
6. We call Y **coisotropic** when $Y^{\Omega} \subseteq Y$.
    Check that every codimension 1 subspace Y is coisotropic.
7. An isotropic subspace Y of (V, Ω) is called **lagrangian** when $\dim Y = \frac{1}{2} \dim V$.
    Check that:

    $$Y \text{ is lagrangian} \iff Y \text{ is isotropic and coisotropic} \iff Y = Y^{\Omega} \ .$$

8. Show that, if Y is a lagrangian subspace of (V, Ω), then any basis $e_1, \ldots, e_n$ of Y can be extended to a symplectic basis $e_1, \ldots, e_n, f_1, \ldots, f_n$ of (V, Ω).

    Hint: Choose f_1 in W^{Ω}, where W is the linear span of $\{e_2, \ldots, e_n\}$.

9. Show that, if Y is a lagrangian subspace, (V, Ω) is symplectomorphic to the space $(Y \oplus Y^*, \Omega_0)$, where Ω_0 is determined by the formula

    $$\Omega_0(u \oplus \alpha, v \oplus \beta) = \beta(u) - \alpha(v) \ .$$

    In fact, for any vector space E, the direct sum $V = E \oplus E^*$ has a canonical symplectic structure determined by the formula above. If $e_1, \ldots, e_n$ is a basis of E, and $f_1, \ldots, f_n$ is the dual basis, then $e_1 \oplus 0, \ldots, e_n \oplus 0, 0 \oplus f_1, \ldots, 0 \oplus f_n$ is a symplectic basis for V.

Chapter 2
Symplectic Form on the Cotangent Bundle

2.1 Cotangent Bundle

Let X be any n-dimensional manifold and $M = T^*X$ its cotangent bundle. If the manifold structure on X is described by coordinate charts $(\mathcal{U}, x_1, \ldots, x_n)$ with $x_i : \mathcal{U} \to \mathbb{R}$, then at any $x \in \mathcal{U}$, the differentials $(dx_1)_x, \ldots (dx_n)_x$ form a basis of T_x^*X. Namely, if $\xi \in T_x^*X$, then $\xi = \sum_{i=1}^n \xi_i (dx_i)_x$ for some real coefficients $\xi_1, \ldots, \xi_n$. This induces a map

$$T^*\mathcal{U} \longrightarrow \mathbb{R}^{2n}$$
$$(x, \xi) \longmapsto (x_1, \ldots, x_n, \xi_1, \ldots, \xi_n) \ .$$

The chart $(T^*\mathcal{U}, x_1, \ldots, x_n, \xi_1, \ldots, \xi_n)$ is a coordinate chart for T^*X; the coordinates $x_1, \ldots, x_n, \xi_1, \ldots, \xi_n$ are the **cotangent coordinates** associated to the coordinates $x_1, \ldots, x_n$ on $\mathcal{U}$. The transition functions on the overlaps are smooth: given two charts $(\mathcal{U}, x_1, \ldots, x_n)$, $(\mathcal{U}', x_1', \ldots, x_n')$, and $x \in \mathcal{U} \cap \mathcal{U}'$, if $\xi \in T_x^*X$, then

$$\xi = \sum_{i=1}^n \xi_i (dx_i)_x = \sum_{i,j} \xi_i \left(\frac{\partial x_i}{\partial x_j'} \right) (dx_j')_x = \sum_{j=1}^n \xi_j' (dx_j')_x$$

where $\xi_j' = \sum_i \xi_i \left(\frac{\partial x_i}{\partial x_j'} \right)$ is smooth. Hence, T^*X is a $2n$-dimensional manifold.

We will now construct a major class of examples of symplectic forms. The *canonical forms* on cotangent bundles are relevant for several branches, including analysis of differential operators, dynamical systems and classical mechanics.

2.2 Tautological and Canonical Forms in Coordinates

Let $(\mathcal{U}, x_1, \ldots, x_n)$ be a coordinate chart for X, with associated cotangent coordinates $(T^*\mathcal{U}, x_1, \ldots, x_n, \xi_1, \ldots, \xi_n)$. Define a 2-form ω on $T^*\mathcal{U}$ by

$$\omega = \sum_{i=1}^{n} dx_i \wedge d\xi_i .$$

In order to check that this definition is coordinate-independent, consider the 1-form on $T^*\mathcal{U}$

$$\alpha = \sum_{i=1}^{n} \xi_i \, dx_i .$$

Clearly, $\omega = -d\alpha$.

Claim. The form α is intrinsically defined (and hence the form ω is also intrinsically defined) .

Proof. Let $(\mathcal{U}, x_1, \ldots, x_n, \xi_1, \ldots, \xi_n)$ and $(\mathcal{U}', x_1', \ldots, x_n', \xi_1', \ldots, \xi_n')$ be two cotangent coordinate charts. On $\mathcal{U} \cap \mathcal{U}'$, the two sets of coordinates are related by $\xi_j' = \sum_i \xi_i \left(\frac{\partial x_i}{\partial x_j'} \right)$. Since $dx_j' = \sum_i \left(\frac{\partial x_j'}{\partial x_i} \right) dx_i$, we have

$$\alpha = \sum_i \xi_i dx_i = \sum_j \xi_j' dx_j' = \alpha' .$$

$\square$

The 1-form α is the **tautological form** or **Liouville 1-form** and the 2-form ω is the **canonical symplectic form**. The following section provides an alternative proof of the intrinsic character of these forms.

2.3 Coordinate-Free Definitions

Let

$$M = T^*X \qquad p = (x, \xi) \qquad \xi \in T_x^*X$$
$$\downarrow \pi \qquad\qquad \downarrow$$
$$X \qquad\qquad x$$

be the natural projection. The **tautological 1-form** α may be defined pointwise by

$$\alpha_p = (d\pi_p)^* \xi \qquad \in T_p^*M ,$$

where $(d\pi_p)^*$ is the transpose of $d\pi_p$, that is, $(d\pi_p)^*\xi = \xi \circ d\pi_p$:

$$
\begin{array}{ccc}
p = (x,\xi) & T_pM & T_p^*M \\
\downarrow \pi & \downarrow d\pi_p & \uparrow (d\pi_p)^* \\
x & T_xX & T_x^*X
\end{array}
$$

Equivalently,

$$\alpha_p(v) = \xi\Big((d\pi_p)v\Big), \quad \text{for } v \in T_pM .$$

Exercise. Let $(\mathcal{U}, x_1, \ldots, x_n)$ be a chart on X with associated cotangent coordinates $x_1, \ldots, x_n, \xi_1, \ldots, \xi_n$. Show that on $T^*\mathcal{U}$, $\alpha = \sum_{i=1}^{n} \xi_i \, dx_i$. ◇

The **canonical symplectic 2-form** ω on T^*X is defined as

$$\omega = -d\alpha .$$

Locally, $\omega = \sum_{i=1}^{n} dx_i \wedge d\xi_i$.

Exercise. Show that the tautological 1-form α is uniquely characterized by the property that, for every 1-form $\mu : X \to T^*X$, $\mu^*\alpha = \mu$. (See Lecture 3.) ◇

2.4 Naturality of the Tautological and Canonical Forms

Let X_1 and X_2 be n-dimensional manifolds with cotangent bundles $M_1 = T^*X_1$ and $M_2 = T^*X_2$, and tautological 1-forms α_1 and α_2. Suppose that $f : X_1 \to X_2$ is a diffeomorphism. Then there is a natural diffeomorphism

$$f_\sharp : M_1 \to M_2$$

which **lifts** f; namely, if $p_1 = (x_1, \xi_1) \in M_1$ for $x_1 \in X_1$ and $\xi_1 \in T_{x_1}^*X_1$, then we define

$$f_\sharp(p_1) = p_2 = (x_2, \xi_2) , \quad \text{with } \begin{cases} x_2 = f(x_1) \in X_2 \text{ and} \\ \xi_1 = (df_{x_1})^*\xi_2 , \end{cases}$$

where $(df_{x_1})^* : T_{x_2}^*X_2 \xrightarrow{\simeq} T_{x_1}^*X_1$, so $f_\sharp|_{T_{x_1}^*}$ is the inverse map of $(df_{x_1})^*$.

Exercise. Check that $f_\sharp$ is a diffeomorphism. Here are some hints:

1.
$$
\begin{array}{ccc}
M_1 & \xrightarrow{f_\sharp} & M_2 \\
\pi_1 \downarrow & & \downarrow \pi_2 \\
X_1 & \xrightarrow{f} & X_2
\end{array}
$$
commutes;

2. $f_\sharp : M_1 \to M_2$ is bijective;
3. $f_\sharp$ and $f_\sharp^{-1}$ are smooth.

Proposition 2.1. *The lift $f_\sharp$ of a diffeomorphism $f : X_1 \to X_2$ pulls the tautological form on T^*X_2 back to the tautological form on T^*X_1, i.e.,*

$$(f_\sharp)^* \alpha_2 = \alpha_1 \ .$$

Proof. At $p_1 = (x_1, \xi_1) \in M_1$, this identity says

$$\left(df_\sharp\right)^*_{p_1} (\alpha_2)_{p_2} = (\alpha_1)_{p_1} \qquad\qquad (\star)$$

where $p_2 = f_\sharp(p_1)$.

Using the following facts,

- Definition of $f_\sharp$:
  $p_2 = f_\sharp(p_1) \iff p_2 = (x_2, \xi_2)$ where $x_2 = f(x_1)$ and $(df_{x_1})^* \xi_2 = \xi_1$.
- Definition of tautological 1-form:
  $(\alpha_1)_{p_1} = (d\pi_1)^*_{p_1} \xi_1$ and $(\alpha_2)_{p_2} = (d\pi_2)^*_{p_2} \xi_2$.

- The diagram $\begin{array}{ccc} M_1 & \xrightarrow{f_\sharp} & M_2 \\ \pi_1 \downarrow & & \downarrow \pi_2 \\ X_1 & \xrightarrow{f} & X_2 \end{array}$ commutes.

the proof of $(\star)$ is:

$$\begin{aligned}
(df_\sharp)^*_{p_1} (\alpha_2)_{p_2} &= (df_\sharp)^*_{p_1} (d\pi_2)^*_{p_2} \xi_2 = \left(d(\pi_2 \circ f_\sharp)\right)^*_{p_1} \xi_2 \\
&= (d(f \circ \pi_1))^*_{p_1} \xi_2 \ = (d\pi_1)^*_{p_1} (df)^*_{x_1} \xi_2 \\
&= (d\pi_1)^*_{p_1} \xi_1 \qquad\qquad = (\alpha_1)_{p_1} \ .
\end{aligned}$$

$\square$

Corollary 2.2. *The lift $f_\sharp$ of a diffeomorphism $f : X_1 \to X_2$ is a symplectomorphism, i.e.,*

$$(f_\sharp)^* \omega_2 = \omega_1 \ ,$$

where ω_1, ω_2 are the canonical symplectic forms.

In summary, a diffeomorphism of manifolds induces a canonical symplectomorphism of cotangent bundles:

$$f_\sharp : T^*X_1 \longrightarrow T^*X_2$$
$$\uparrow$$
$$f : \ X_1 \ \longrightarrow \ X_2$$

Example. Let $X_1 = X_2 = S^1$. Then T^*S^1 is an infinite cylinder $S^1 \times \mathbb{R}$. The canonical 2-form ω is the area form $\omega = d\theta \wedge d\xi$. If $f : S^1 \to S^1$ is any diffeomorphism, then $f_\sharp : S^1 \times \mathbb{R} \to S^1 \times \mathbb{R}$ is a symplectomorphism, i.e., is an area-preserving diffeomorphism of the cylinder. $\diamondsuit$

If $f : X_1 \to X_2$ and $g : X_2 \to X_3$ are diffeomorphisms, then $(g \circ f)_\sharp = g_\sharp \circ f_\sharp$. In terms of the group $\mathrm{Diff}(X)$ of diffeomorphisms of X and the group $\mathrm{Sympl}(M, \omega)$ of symplectomorphisms of (M, ω), we say that the map

$$\mathrm{Diff}(X) \longrightarrow \mathrm{Sympl}(M, \omega)$$
$$f \longmapsto f_\sharp$$

is a group homomorphism. This map is clearly injective. Is it surjective? Do all symplectomorphisms $T^*X \to T^*X$ come from diffeomorphisms $X \to X$? No: for instance, translation along cotangent fibers is not induced by a diffeomorphism of the base manifold. A criterion for which symplectomorphisms arise as lifts of diffeomorphisms is discussed in Homework 3.

Homework 2: Symplectic Volume

1. Given a vector space V, the exterior algebra of its dual space is

$$\wedge^*(V^*) = \bigoplus_{k=0}^{\dim V} \wedge^k(V^*),$$

where $\wedge^k(V^*)$ is the set of maps $\alpha : \overbrace{V \times \cdots \times V}^{k} \to \mathbb{R}$ which are linear in each entry, and for any permutation π, $\alpha(v_{\pi_1}, \ldots, v_{\pi_k}) = (\text{sign } \pi) \cdot \alpha(v_1, \ldots, v_k)$. The elements of $\wedge^k(V^*)$ are known as **skew-symmetric k-linear maps** or **k-forms** on V.

   (a) Show that any $\Omega \in \wedge^2(V^*)$ is of the form $\Omega = e_1^* \wedge f_1^* + \ldots + e_n^* \wedge f_n^*$, where $u_1^*, \ldots, u_k^*, e_1^*, \ldots, e_n^*, f_1^*, \ldots, f_n^*$ is a basis of V^* dual to the standard basis $(k + 2n = \dim V)$.

   (b) In this language, a symplectic map $\Omega : V \times V \to \mathbb{R}$ is just a nondegenerate 2-form $\Omega \in \wedge^2(V^*)$, called a **symplectic form** on V.

   Show that, if Ω is any symplectic form on a vector space V of dimension $2n$, then the nth exterior power $\Omega^n = \underbrace{\Omega \wedge \ldots \wedge \Omega}_{n}$ does not vanish.

   (c) Deduce that the nth exterior power ω^n of any symplectic form ω on a $2n$-dimensional manifold M is a volume form.[1]

   Hence, any symplectic manifold (M, ω) is canonically oriented by the symplectic structure. The form $\frac{\omega^n}{n!}$ is called the **symplectic volume** or the **Liouville volume** of (M, ω).

   Does the Möbius strip support a symplectic structure?

   (d) Conversely, given a 2-form $\Omega \in \wedge^2(V^*)$, show that, if $\Omega^n \neq 0$, then Ω is symplectic.

   Hint: Standard form.

2. Let (M, ω) be a $2n$-dimensional symplectic manifold, and let ω^n be the volume form obtained by wedging ω with itself n times.

   (a) Show that, if M is compact, the de Rham cohomology class $[\omega^n] \in H^{2n}(M; \mathbb{R})$ is non-zero.

   Hint: Stokes' theorem.

   (b) Conclude that $[\omega]$ itself is non-zero (in other words, that ω is not exact).
   (c) Show that if $n > 1$ there are no symplectic structures on the sphere S^{2n}.

---

[1] A **volume form** is a nonvanishing form of top degree.

Part II
Symplectomorphisms

Equivalence between symplectic manifolds is expressed by a *symplectomorphism*. By Weinstein's lagrangian creed [105], everything is a lagrangian manifold! We will study symplectomorphisms according to the creed.

Chapter 3
Lagrangian Submanifolds

3.1 Submanifolds

Let M and X be manifolds with $\dim X < \dim M$.

Definition 3.1. A map $i : X \to M$ is an ***immersion*** if $di_p : T_pX \to T_{i(p)}M$ is injective for any point $p \in X$.

An ***embedding*** is an immersion which is a homeomorphism onto its image.[1]

A ***closed embedding*** is a proper[2] injective immersion.

Exercise. Show that a map $i : X \to M$ is a closed embedding if and only if i is an embedding and its image $i(X)$ is closed in M.

Hints:

- If i is injective and proper, then for any neighborhood $\mathcal{U}$ of $p \in X$, there is a neighborhood $\mathcal{V}$ of $i(p)$ such that $f^{-1}(\mathcal{V}) \subseteq \mathcal{U}$.
- On a Hausdorff space, any compact set is closed. On any topological space, a closed subset of a compact set is compact.
- An embedding is proper if and only if its image is closed.

Definition 3.2. A ***submanifold*** of M is a manifold X with a closed embedding $i : X \hookrightarrow M$.[3]

Notation. Given a submanifold, we regard the embedding $i : X \hookrightarrow M$ as an inclusion, in order to identify points and tangent vectors:

$$p = i(p) \quad \text{and} \quad T_pX = di_p(T_pX) \subset T_pM .$$

---

[1] The image has the topology induced by the target manifold.

[2] A map is **proper** if the preimage of any compact set is compact.

[3] When X is an open subset of a manifold M, we refer to it as an *open* submanifold.

3.2 Lagrangian Submanifolds of T^*X

Definition 3.3. Let (M, ω) be a $2n$-dimensional symplectic manifold. A submanifold Y of M is a **lagrangian submanifold** if, at each $p \in Y$, T_pY is a lagrangian subspace of T_pM, i.e., $\omega_p|_{T_pY} \equiv 0$ and $\dim T_pY = \frac{1}{2} \dim T_pM$. Equivalently, if $i : Y \hookrightarrow M$ is the inclusion map, then Y is **lagrangian** if and only if $i^*\omega = 0$ and $\dim Y = \frac{1}{2} \dim M$.

Let X be an n-dimensional manifold, with $M = T^*X$ its cotangent bundle. If $x_1, \ldots, x_n$ are coordinates on $U \subseteq X$, with associated cotangent coordinates $x_1, \ldots, x_n, \xi_1, \ldots, \xi_n$ on T^*U, then the tautological 1-form on T^*X is

$$\alpha = \sum \xi_i dx_i$$

and the canonical 2-form on T^*X is

$$\omega = -d\alpha = \sum dx_i \wedge d\xi_i \ .$$

The **zero section** of T^*X

$$X_0 := \{(x, \xi) \in T^*X \mid \xi = 0 \text{ in } T_x^*X\}$$

is an n-dimensional submanifold of T^*X whose intersection with T^*U is given by the equations $\xi_1 = \cdots = \xi_n = 0$. Clearly $\alpha = \sum \xi_i dx_i$ vanishes on $X_0 \cap T^*U$. In particular, if $i_0 : X_0 \hookrightarrow T^*X$ is the inclusion map, we have $i_0^*\alpha = 0$. Hence, $i_0^*\omega = i_0^*d\alpha = 0$, and X_0 is lagrangian.

What are all the lagrangian submanifolds of T^*X which are "C^1-close to X_0"?

Let X_μ be (the image of) another section, that is, an n-dimensional submanifold of T^*X of the form

$$X_\mu = \{(x, \mu_x) \mid x \in X, \ \mu_x \in T_x^*X\} \qquad (\star)$$

where the covector μ_x depends smoothly on x, and $\mu : X \to T^*X$ is a de Rham 1-form. Relative to the inclusion $i : X_\mu \hookrightarrow T^*X$ and the cotangent projection $\pi : T^*X \to X$, X_μ is of the form $(\star)$ if and only if $\pi \circ i : X_\mu \to X$ is a diffeomorphism.

When is such an X_μ lagrangian?

Proposition 3.4. *Let X_μ be of the form $(\star)$, and let μ be the associated de Rham 1-form. Denote by $s_\mu : X \to T^*X$, $x \mapsto (x, \mu_x)$, the 1-form μ regarded exclusively as a map. Notice that the image of s_μ is X_μ. Let α be the tautological 1-form on T^*X. Then*

$$s_\mu^*\alpha = \mu \ .$$

Proof. By definition of α (previous lecture), $\alpha_p = (d\pi_p)^*\xi$ at $p = (x, \xi) \in M$. For $p = s_\mu(x) = (x, \mu_x)$, we have $\alpha_p = (d\pi_p)^*\mu_x$. Then

$$(s_\mu^*\alpha)_x = (ds_\mu)_x^*\alpha_p = (ds_\mu)_x^*(d\pi_p)^*\mu_x = (d(\underbrace{\pi \circ s_\mu}_{\text{id}_X}))_x^*\mu_x = \mu_x \ .$$

$\square$

Suppose that X_μ is an n-dimensional submanifold of T^*X of the form $(\star)$, with associated de Rham 1-form μ. Then $s_\mu : X \to T^*X$ is an embedding with image X_μ, and there is a diffeomorphism $\tau : X \to X_\mu$, $\tau(x) := (x, \mu_x)$, such that the following diagram commutes.

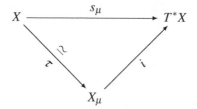

We want to express the condition of X_μ being lagrangian in terms of the form μ:

$$
\begin{aligned}
X_\mu \text{ is lagrangian } &\Longleftrightarrow i^* d\alpha = 0 \\
&\Longleftrightarrow \tau^* i^* d\alpha = 0 \\
&\Longleftrightarrow (i \circ \tau)^* d\alpha = 0 \\
&\Longleftrightarrow s_\mu^* d\alpha = 0 \\
&\Longleftrightarrow d s_\mu^* \alpha = 0 \\
&\Longleftrightarrow d\mu = 0 \\
&\Longleftrightarrow \mu \text{ is closed .}
\end{aligned}
$$

Therefore, there is a one-to-one correspondence between the set of lagrangian submanifolds of T^*X of the form $(\star)$ and the set of closed 1-forms on X.

When X is simply connected, $H^1_{\text{deRham}}(X) = 0$, so every closed 1-form μ is equal to df for some $f \in C^\infty(X)$. Any such primitive f is then called a **generating function** for the lagrangian submanifold X_μ associated to μ. (Two functions generate the same lagrangian submanifold if and only if they differ by a locally constant function.) On arbitrary manifolds X, functions $f \in C^\infty(X)$ originate lagrangian submanifolds as images of df.

Exercise. Check that, if X is compact (and not just one point) and $f \in C^\infty(X)$, then $\#(X_{df} \cap X_0) \geq 2$. $\Diamond$

There are lots of lagrangian submanifolds of T^*X not covered by the description in terms of closed 1-forms, starting with the cotangent fibers.

3.3 Conormal Bundles

Let S be any k-dimensional submanifold of an n-dimensional manifold X.

Definition 3.5. The *conormal space* at $x \in S$ is

$$
N_x^* S = \{ \xi \in T_x^* X \mid \xi(v) = 0 \text{ , for all } v \in T_x S \} .
$$

The *conormal bundle* of S is

$$N^*S = \{(x,\xi) \in T^*X \mid x \in S,\ \xi \in N_x^*S\} .$$

Exercise. The conormal bundle N^*S is an n-dimensional submanifold of T^*X.
    Hint: Use coordinates on X adapted[4] to S.

Proposition 3.6. *Let* $i : N^*S \hookrightarrow T^*X$ *be the inclusion, and let* α *be the tautological 1-form on* T^*X. *Then*

$$i^*\alpha = 0 .$$

Proof. Let $(\mathcal{U}, x_1, \ldots, x_n)$ be a coordinate system on X centered at $x \in S$ and adapted to S, so that $\mathcal{U} \cap S$ is described by $x_{k+1} = \cdots = x_n = 0$. Let $(T^*\mathcal{U}, x_1, \ldots, x_n, \xi_1, \ldots, \xi_n)$ be the associated cotangent coordinate system. The submanifold $N^*S \cap T^*\mathcal{U}$ is then described by

$$x_{k+1} = \cdots = x_n = 0 \qquad \text{and} \qquad \xi_1 = \cdots = \xi_k = 0 .$$

Since $\alpha = \sum \xi_i dx_i$ on $T^*\mathcal{U}$, we conclude that, at $p \in N^*S$,

$$(i^*\alpha)_p = \alpha_p|_{T_p(N^*S)} = \left. \sum_{i>k} \xi_i dx_i \right|_{\mathrm{span}\{\frac{\partial}{\partial x_i},\, i \leq k\}} = 0 .$$

$\square$

Corollary 3.7. *For any submanifold* $S \subset X$, *the conormal bundle* N^*S *is a lagrangian submanifold of* T^*X.

Taking $S = \{x\}$ to be one point, the conormal bundle $L = N^*S = T_x^*X$ is a cotangent fiber. Taking $S = X$, the conormal bundle $L = X_0$ is the zero section of T^*X.

3.4 Application to Symplectomorphisms

Let (M_1, ω_1) and (M_2, ω_2) be two $2n$-dimensional symplectic manifolds. Given a diffeomorphism $\varphi : M_1 \xrightarrow{\simeq} M_2$, when is it a symplectomorphism? (I.e., when is $\varphi^*\omega_2 = \omega_1$?)

---

[4] A coordinate chart $(\mathcal{U}, x_1, \ldots, x_n)$ on X is adapted to a k-dimensional submanifold S if $S \cap \mathcal{U}$ is described by $x_{k+1} = \cdots = x_n = 0$.

Consider the two projection maps

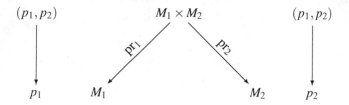

Then $\omega = (\mathrm{pr}_1)^*\omega_1 + (\mathrm{pr}_2)^*\omega_2$ is a 2-form on $M_1 \times M_2$ which is closed,

$$d\omega = (\mathrm{pr}_1)^* \underbrace{d\omega_1}_{0} + (\mathrm{pr}_2)^* \underbrace{d\omega_2}_{0} = 0 \;,$$

and symplectic,

$$\omega^{2n} = \binom{2n}{n} \left((\mathrm{pr}_1)^*\omega_1\right)^n \wedge \left((\mathrm{pr}_2)^*\omega_2\right)^n \neq 0 \;.$$

More generally, if $\lambda_1, \lambda_2 \in \mathbb{R}\setminus\{0\}$, then $\lambda_1(\mathrm{pr}_1)^*\omega_1 + \lambda_2(\mathrm{pr}_2)^*\omega_2$ is also a symplectic form on $M_1 \times M_2$. Take $\lambda_1 = 1$, $\lambda_2 = -1$ to obtain the **twisted product form** on $M_1 \times M_2$:

$$\widetilde{\omega} = (\mathrm{pr}_1)^*\omega_1 - (\mathrm{pr}_2)^*\omega_2 \;.$$

The graph of a diffeomorphism $\varphi : M_1 \xrightarrow{\sim} M_2$ is the $2n$-dimensional submanifold of $M_1 \times M_2$:

$$\Gamma_\varphi := \mathrm{Graph}\,\varphi = \{(p, \varphi(p)) \mid p \in M_1\} \;.$$

The submanifold Γ_φ is an embedded image of M_1 in $M_1 \times M_2$, the embedding being the map

$$\gamma : M_1 \longrightarrow M_1 \times M_2$$
$$p \longmapsto (p, \varphi(p)) \;.$$

Proposition 3.8. *A diffeomorphism φ is a symplectomorphism if and only if Γ_φ is a lagrangian submanifold of $(M_1 \times M_2, \widetilde{\omega})$.*

Proof. The graph Γ_φ is lagrangian if and only if $\gamma^*\widetilde{\omega} = 0$. But

$$\gamma^*\widetilde{\omega} = \gamma^*\,\mathrm{pr}_1^*\,\omega_1 - \gamma^*\,\mathrm{pr}_2^*\,\omega_2$$
$$= (\mathrm{pr}_1 \circ \gamma)^*\omega_1 - (\mathrm{pr}_2 \circ \gamma)^*\omega_2$$

and $\mathrm{pr}_1 \circ \gamma$ is the identity map on M_1 whereas $\mathrm{pr}_2 \circ \gamma = \varphi$. Therefore,

$$\gamma^*\widetilde{\omega} = 0 \quad \Longleftrightarrow \quad \varphi^*\omega_2 = \omega_1 \;.$$

$\square$

Homework 3:
Tautological Form and Symplectomorphisms

This set of problems is from [53].

1. Let (M, ω) be a symplectic manifold, and let α be a 1-form such that

$$\omega = -d\alpha .$$

Show that there exists a unique vector field v such that its interior product with ω is α, i.e., $\iota_v \omega = -\alpha$.
Prove that, if g is a symplectomorphism which preserves α (that is, $g^* \alpha = \alpha$), then g commutes with the one-parameter group of diffeomorphisms generated by v, i.e.,

$$(\exp tv) \circ g = g \circ (\exp tv) .$$

Hint: Recall that, for $p \in M$, $(\exp tv)(p)$ is the *unique* curve in M solving the ordinary differential equation

$$\begin{cases} \frac{d}{dt}(\exp tv(p)) = v(\exp tv(p)) \\ (\exp tv)(p)|_{t=0} = p \end{cases}$$

for t in some neighborhood of 0. Show that $g \circ (\exp tv) \circ g^{-1}$ is the one-parameter group of diffeomorphisms generated by $g_* v$. (The push-forward of v by g is defined by $(g_* v)_{g(p)} = dg_p(v_p)$.) Finally check that g preserves v (that is, $g_* v = v$).

2. Let X be an arbitrary n-dimensional manifold, and let $M = T^* X$. Let $(\mathcal{U}, x_1, \ldots, x_n)$ be a coordinate system on X, and let $x_1, \ldots, x_n, \xi_1, \ldots, \xi_n$ be the corresponding coordinates on $T^* \mathcal{U}$.
Show that, when α is the tautological 1-form on M (which, in these coordinates, is $\sum \xi_i \, dx_i$), the vector field v in the previous exercise is just the vector field $\sum \xi_i \frac{\partial}{\partial \xi_i}$.
Let $\exp tv$, $-\infty < t < \infty$, be the one-parameter group of diffeomorphisms generated by v.
Show that, for every point $p = (x, \xi)$ in M,

$$(\exp tv)(p) = p_t \quad \text{where} \quad p_t = (x, e^t \xi) .$$

3. Let M be as in exercise 2.
Show that, if g is a symplectomorphism of M which preserves α, then

$$g(x, \xi) = (y, \eta) \quad \Longrightarrow \quad g(x, \lambda \xi) = (y, \lambda \eta)$$

for all $(x, \xi) \in M$ and $\lambda \in \mathbb{R}$.
Conclude that g has to preserve the cotangent fibration, i.e., show that there exists a diffeomorphism $f : X \to X$ such that $\pi \circ g = f \circ \pi$, where $\pi : M \to X$ is the projection map $\pi(x, \xi) = x$.
Finally prove that $g = f_\#$, the map $f_\#$ being the symplectomorphism of M lifting f.

Hint: Suppose that $g(p) = q$ where $p = (x, \xi)$ and $q = (y, \eta)$. Combine the identity

$$(dg_p)^* \alpha_q = \alpha_p$$

with the identity

$$d\pi_q \circ dg_p = df_x \circ d\pi_p .$$

(The first identity expresses the fact that $g^* \alpha = \alpha$, and the second identity is obtained by differentiating both sides of the equation $\pi \circ g = f \circ \pi$ at p.)

4. Let M be as in exercise 2, and let h be a smooth function on X. Define $\tau_h : M \to M$ by setting

$$\tau_h(x, \xi) = (x, \xi + dh_x) .$$

Prove that

$$\tau_h^* \alpha = \alpha + \pi^* dh$$

where π is the projection map

$$
\begin{array}{cc}
M & (x, \xi) \\
\downarrow \pi & \downarrow \\
X & x
\end{array}
$$

Deduce that

$$\tau_h^* \omega = \omega ,$$

i.e., that τ_h is a symplectomorphism.

Chapter 4
Generating Functions

4.1 Constructing Symplectomorphisms

Let X_1, X_2 be n-dimensional manifolds, with cotangent bundles $M_1 = T^*X_1$, $M_2 = T^*X_2$, tautological 1-forms α_1, α_2, and canonical 2-forms ω_1, ω_2.

Under the natural identification

$$M_1 \times M_2 = T^*X_1 \times T^*X_2 \simeq T^*(X_1 \times X_2) ,$$

the tautological 1-form on $T^*(X_1 \times X_2)$ is

$$\alpha = (\mathrm{pr}_1)^* \alpha_1 + (\mathrm{pr}_2)^* \alpha_2 ,$$

where $\mathrm{pr}_i : M_1 \times M_2 \to M_i$, $i = 1, 2$ are the two projections. The canonical 2-form on $T^*(X_1 \times X_2)$ is

$$\omega = -d\alpha = -d\mathrm{pr}_1^* \alpha_1 - d\mathrm{pr}_2^* \alpha_2 = \mathrm{pr}_1^* \omega_1 + \mathrm{pr}_2^* \omega_2 .$$

In order to describe the twisted form $\widetilde{\omega} = \mathrm{pr}_1^* \omega_1 - \mathrm{pr}_2^* \omega_2$, we define an involution of $M_2 = T^*X_2$ by

$$\sigma_2 : \quad \begin{array}{ccc} M_2 & \longrightarrow & M_2 \\ (x_2, \xi_2) & \longmapsto & (x_2, -\xi_2) \end{array}$$

which yields $\sigma_2^* \alpha_2 = -\alpha_2$. Let $\sigma = \mathrm{id}_{M_1} \times \sigma_2 : M_1 \times M_2 \to M_1 \times M_2$. Then

$$\sigma^* \widetilde{\omega} = \mathrm{pr}_1^* \omega_1 + \mathrm{pr}_2^* \omega_2 = \omega .$$

If Y is a lagrangian submanifold of $(M_1 \times M_2, \omega)$, then its "twist" $Y^\sigma := \sigma(Y)$ is a lagrangian submanifold of $(M_1 \times M_2, \widetilde{\omega})$.

Recipe for producing symplectomorphisms $M_1 = T^*X_1 \to M_2 = T^*X_2$:

1. Start with a lagrangian submanifold Y of $(M_1 \times M_2, \omega)$.
2. Twist it to obtain a lagrangian submanifold Y^σ of $(M_1 \times M_2, \widetilde{\omega})$.

3. Check whether Y^σ is the graph of some diffeomorphism $\varphi : M_1 \to M_2$.
4. If it is, then φ is a symplectomorphism (by Proposition 3.8).

Let $i : Y^\sigma \hookrightarrow M_1 \times M_2$ be the inclusion map

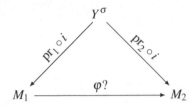

Step 3 amounts to checking whether $\mathrm{pr}_1 \circ i$ and $\mathrm{pr}_2 \circ i$ are diffeomorphisms. If yes, then $\varphi := (\mathrm{pr}_2 \circ i) \circ (\mathrm{pr}_1 \circ i)^{-1}$ is a diffeomorphism.

In order to obtain lagrangian submanifolds of $M_1 \times M_2 \simeq T^*(X_1 \times X_2)$, we can use the *method of generating functions*.

4.2 Method of Generating Functions

For any $f \in C^\infty(X_1 \times X_2)$, df is a closed 1-form on $X_1 \times X_2$. The **lagrangian submanifold generated by** f is

$$Y_f := \{((x,y),(df)_{(x,y)}) \mid (x,y) \in X_1 \times X_2\} .$$

We adopt the notation

$$d_x f := (df)_{(x,y)} \text{ projected to } T^*_x X_1 \times \{0\},$$
$$d_y f := (df)_{(x,y)} \text{ projected to } \{0\} \times T^*_y X_2 ,$$

which enables us to write

$$Y_f = \{(x,y,d_x f, d_y f) \mid (x,y) \in X_1 \times X_2\}$$

and

$$Y_f^\sigma = \{(x,y,d_x f, -d_y f) \mid (x,y) \in X_1 \times X_2\} .$$

When Y_f^σ is in fact the graph of a diffeomorphism $\varphi : M_1 \to M_2$, we call φ the **symplectomorphism generated by** f, and call f the **generating function**, of $\varphi : M_1 \to M_2$.

So when is Y_f^σ the graph of a diffeomorphism $\varphi : M_1 \to M_2$?

Let $(\mathcal{U}_1, x_1, \ldots, x_n), (\mathcal{U}_2, y_1, \ldots, y_n)$ be coordinate charts for X_1, X_2, with associated charts $(T^*\mathcal{U}_1, x_1, \ldots, x_n, \xi_1, \ldots, \xi_n)$, $(T^*\mathcal{U}_2, y_1, \ldots, y_n, \eta_1, \ldots, \eta_n)$ for M_1, M_2. The set Y_f^σ is the graph of $\varphi : M_1 \to M_2$ if and only if, for any $(x, \xi) \in M_1$ and $(y, \eta) \in M_2$, we have

$$\varphi(x,\xi) = (y,\eta) \iff \xi = d_x f \text{ and } \eta = -d_y f.$$

Therefore, given a point $(x,\xi) \in M_1$, to find its image $(y,\eta) = \varphi(x,\xi)$ we must solve the "Hamilton" equations

$$
\begin{cases}
\xi_i = \dfrac{\partial f}{\partial x_i}(x,y) & (\star) \\[2mm]
\eta_i = -\dfrac{\partial f}{\partial y_i}(x,y). & (\star\star)
\end{cases}
$$

If there is a solution $y = \varphi_1(x,\xi)$ of $(\star)$, we may feed it to $(\star\star)$ thus obtaining $\eta = \varphi_2(x,\xi)$, so that $\varphi(x,\xi) = (\varphi_1(x,\xi), \varphi_2(x,\xi))$. Now by the implicit function theorem, in order to solve $(\star)$ locally for y in terms of x and ξ, we need the condition

$$\det\left[\frac{\partial}{\partial y_j}\left(\frac{\partial f}{\partial x_i}\right)\right]^n_{i,j=1} \neq 0.$$

This is a necessary local condition for f to generate a symplectomorphism φ. Locally this is also sufficient, but globally there is the usual bijectivity issue.

Example. Let $X_1 = \mathcal{U}_1 \simeq \mathbb{R}^n$, $X_2 = \mathcal{U}_2 \simeq \mathbb{R}^n$, and $f(x,y) = -\frac{|x-y|^2}{2}$, the square of euclidean distance up to a constant.

The "Hamilton" equations are

$$
\begin{cases}
\xi_i = \dfrac{\partial f}{\partial x_i} = y_i - x_i \\[2mm]
\eta_i = -\dfrac{\partial f}{\partial y_i} - y_l - x_i
\end{cases}
\iff
\begin{cases}
y_i = x_i + \xi_i \\[2mm]
\eta_i = \xi_i.
\end{cases}
$$

The symplectomorphism generated by f is

$$\varphi(x,\xi) = (x+\xi,\xi).$$

If we use the euclidean inner product to identify $T^*\mathbb{R}^n$ with $T\mathbb{R}^n$, and hence regard φ as $\widetilde{\varphi} : T\mathbb{R}^n \to T\mathbb{R}^n$ and interpret ξ as the velocity vector, then the symplectomorphism φ corresponds to free translational motion in euclidean space.

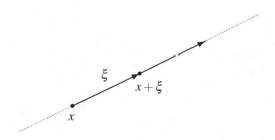

4.3 Application to Geodesic Flow

Let V be an n-dimensional vector space. A **positive inner product** G on V is a bilinear map $G : V \times V \to \mathbb{R}$ which is

symmetric : $G(v,w) = G(w,v)$,   and
positive-definite : $G(v,v) > 0$ when $v \neq 0$.

Definition 4.1. A *riemannian metric* on a manifold X is a function g which assigns to each point $x \in X$ a positive inner product g_x on $T_x X$.

A riemannian metric g is *smooth* if for every smooth vector field $v : X \to TX$ the real-valued function $x \mapsto g_x(v_x, v_x)$ is a smooth function on X.

Definition 4.2. A *riemannian manifold* (X, g) is a manifold X equipped with a smooth riemannian metric g.

The **arc-length** of a piecewise smooth curve $\gamma : [a, b] \to X$ on a riemannian manifold (X, g) is

$$\int_a^b \sqrt{g_{\gamma(t)}\left(\frac{d\gamma}{dt}, \frac{d\gamma}{dt}\right)} \, dt \ .$$

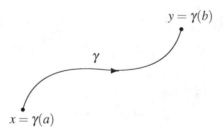

Definition 4.3. The *riemannian distance* between two points x and y of a connected riemannian manifold (X, g) is the infimum $d(x, y)$ of the set of all arc-lengths for piecewise smooth curves joining x to y.

A smooth curve joining x to y is a *minimizing geodesic*[1] if its arc-length is the riemannian distance $d(x, y)$.

A riemannian manifold (X, g) is *geodesically convex* if every point x is joined to every other point y by a unique minimizing geodesic.

Example. On $X = \mathbb{R}^n$ with $TX \simeq \mathbb{R}^n \times \mathbb{R}^n$, let $g_x(v, w) = \langle v, w \rangle$, $g_x(v, v) = |v|^2$, where $\langle \cdot, \cdot \rangle$ is the euclidean inner product, and $|\cdot|$ is the euclidean norm. Then

---

[1] In riemannian geometry, a **geodesic** is a curve which locally minimizes distance and whose velocity is constant.

$(\mathbb{R}^n, \langle \cdot, \cdot \rangle)$ is a geodesically convex riemannian manifold, and the riemannian distance is the usual euclidean distance $d(x,y) = |x-y|$. $\diamondsuit$

Suppose that (X,g) is a geodesically convex riemannian manifold. Consider the function
$$f : X \times X \longrightarrow \mathbb{R}, \qquad f(x,y) = -\frac{d(x,y)^2}{2}.$$

What is the symplectomorphism $\varphi : T^*X \to T^*X$ generated by f?

The metric $g_x : T_xX \times T_xX \to \mathbb{R}$ induces an identification
$$\widetilde{g}_x : T_xX \xrightarrow{\;\simeq\;} T_x^*X$$
$$v \longmapsto g_x(v, \cdot)$$

Use $\widetilde{g}$ to translate φ into a map $\widetilde{\varphi} : TX \to TX$.

We need to solve
$$\begin{cases} \widetilde{g}_x(v) = \xi_i = d_x f(x,y) \\ \widetilde{g}_y(w) = \eta_i = -d_y f(x,y) \end{cases}$$

for (y,η) in terms of (x,ξ) in order to find φ, or, equivalently, for (y,w) in terms (x,v) in order to find $\widetilde{\varphi}$.

Let γ be the geodesic with initial conditions $\gamma(0) = x$ and $\frac{d\gamma}{dt}(0) = v$.

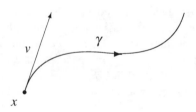

Then the symplectomorphism φ corresponds to the map
$$\widetilde{\varphi} : \quad TX \longrightarrow TX$$
$$(x,v) \longmapsto (\gamma(1), \tfrac{d\gamma}{dt}(1)) .$$

This is called the **geodesic flow** on X (see Homework 4).

Homework 4: Geodesic Flow

This set of problems is adapted from [53].

Let (X, g) be a riemannian manifold. The arc-length of a smooth curve γ:
$[a, b] \to X$ is

$$\text{arc-length of } \gamma := \int_a^b \left| \frac{d\gamma}{dt} \right| dt, \quad \text{where} \quad \left| \frac{d\gamma}{dt} \right| := \sqrt{g_{\gamma(t)} \left(\frac{d\gamma}{dt}, \frac{d\gamma}{dt} \right)}.$$

1. Show that the arc-length of γ is independent of the parametrization of γ, i.e.,
   show that, if we reparametrize γ by $\tau : [a', b'] \to [a, b]$, the new curve $\gamma' = \gamma \circ \tau$:
   $[a', b'] \to X$ has the same arc-length. A curve γ is called a curve of *constant
   velocity* when $\left| \frac{d\gamma}{dt} \right|$ is independent of t. Show that, given any curve $\gamma : [a, b] \to X$
   (with $\frac{d\gamma}{dt}$ never vanishing), there is a reparametrization $\tau : [a, b] \to [a, b]$ such that
   $\gamma \circ \tau : [a, b] \to X$ is of constant velocity.

2. Given a smooth curve $\gamma : [a, b] \to X$, the *action* of γ is $\mathcal{A}(\gamma) := \int_a^b \left| \frac{d\gamma}{dt} \right|^2 dt$.

   Show that, among all curves joining x to y, γ minimizes the action if and only if
   γ is of constant velocity and γ minimizes arc-length.

   Hint: Suppose that γ is of constant velocity, and let $\tau : [a, b] \to [a, b]$ be a reparametrization. Show that $\mathcal{A}(\gamma \circ \tau) \geq \mathcal{A}(\gamma)$, with equality only when $\tau = $ identity.

3. Assume that (X, g) is geodesically convex, that is, any two points $x, y \in X$ are
   joined by a unique (up to reparametrization) minimizing geodesic; its arc-length
   $d(x, y)$ is called the riemannian distance between x and y.
   Assume also that (X, g) is *geodesically complete*, that is, every geodesic can be
   extended indefinitely. Given $(x, v) \in TX$, let $\exp(x, v) : \mathbb{R} \to X$ be the unique
   minimizing geodesic of constant velocity with initial conditions $\exp(x, v)(0) = x$
   and $\frac{d\exp(x,v)}{dt}(0) = v$.
   Consider the function $f : X \times X \to \mathbb{R}$ given by $f(x, y) = -\frac{1}{2} \cdot d(x, y)^2$. Let $d_x f$
   and $d_y f$ be the components of $df_{(x,y)}$ with respect to $T^*_{(x,y)}(X \times X) \simeq T^*_x X \times T^*_y X$.
   Recall that, if

   $$\Gamma_f^\sigma = \{ (x, y, d_x f, -d_y f) \mid (x, y) \in X \times X \}$$

   is the graph of a diffeomorphism $f : T^*X \to T^*X$, then f is the symplectomorphism generated by f. In this case, $f(x, \xi) = (y, \eta)$ if and only if $\xi = d_x f$ and
   $\eta = -d_y f$.
   Show that, under the identification of TX with T^*X by g, the symplectomorphism
   generated by f coincides with the map $TX \to TX$, $(x, v) \mapsto \exp(x, v)(1)$.

Hint: The metric g provides the identifications $T_xXv \simeq \xi(\cdot) = g_x(v, \cdot) \in T_x^*X$. We need to show that, given $(x, v) \in TX$, the unique solution of

$(\star) \begin{cases} g_x(v, \cdot) = d_x f(\cdot) \\ g_y(w, \cdot) = -d_y f(\cdot) \end{cases}$ is $(y, w) = (\exp(x, v)(1), d\frac{\exp(x,v)}{dt}(1))$.

Look up the Gauss lemma in a book on riemannian geometry. It asserts that geodesics are orthogonal to the level sets of the distance function.

To solve the <u>first</u> line in $(\star)$ for y, evaluate both sides at $v = \frac{d\exp(x,v)}{dt}(0)$. Conclude that $y = \exp(x, v)(1)$. Check that $d_x f(v') = 0$ for vectors $v' \in T_xX$ orthogonal to v (that is, $g_x(v, v') = 0$); this is a consequence of $f(x, y)$ being the arc-length of a *minimizing* geodesic, and it suffices to check locally.

The vector w is obtained from the <u>second</u> line of $(\star)$. Compute $-d_y f(\frac{d\exp(x,v)}{dt}(1))$. Then evaluate $-d_y f$ at vectors $w' \in T_yX$ orthogonal to $\frac{d\exp(x,v)}{dt}(1)$; this pairing is again 0 because $f(x, y)$ is the arc-length of a minimizing geodesic. Conclude, using the nondegeneracy of g, that $w = \frac{d\exp(x,v)}{dt}(1)$.

For both steps, it might be useful to recall that, given a function $\varphi : X \to \mathbb{R}$ and a tangent vector $v \in T_xX$, we have $d\varphi_x(v) = \frac{d}{du}[\varphi(\exp(x, v)(u))]_{u=0}$.

Chapter 5
Recurrence

5.1 Periodic Points

Let X be an n-dimensional manifold. Let $M = T^*X$ be its cotangent bundle with canonical symplectic form ω.

Suppose that we are given a smooth function $f : X \times X \to \mathbb{R}$ which generates a symplectomorphism $\varphi : M \to M$, $\varphi(x, d_x f) = (y, -d_y f)$, by the recipe of the previous lecture.

What are the fixed points of φ?

Define $\psi : X \to \mathbb{R}$ by $\psi(x) = f(x, x)$.

Proposition 5.1. *There is a one-to-one correspondence between the fixed points of φ and the critical points of ψ.*

Proof. At $x_0 \in X$, $d_{x_0} \psi = (d_x f + d_y f)|_{(x,y)=(x_0,x_0)}$. Let $\xi = d_x f|_{(x,y)=(x_0,x_0)}$.

$$x_0 \text{ is a critical point of } \psi \iff d_{x_0} \psi = 0 \iff d_y f|_{(x,y)=(x_0,x_0)} = -\xi \ .$$

Hence, the point in Γ_f^σ corresponding to $(x, y) = (x_0, x_0)$ is (x_0, x_0, ξ, ξ). But Γ_f^σ is the graph of φ, so $\varphi(x_0, \xi) = (x_0, \xi)$ is a fixed point. This argument also works backwards. $\qquad \square$

Consider the iterates of φ,

$$\varphi^{(N)} = \underbrace{\varphi \circ \varphi \circ \ldots \circ \varphi}_{N} : M \longrightarrow M \ , \qquad N - 1, 2, \ldots \ ,$$

each of which is a symplectomorphism of M. According to the previous proposition, if $\varphi^{(N)} : M \to M$ is generated by $f^{(N)}$, then there is a correspondence

$$\left\{ \text{fixed points of } \varphi^{(N)} \right\} \xleftrightarrow{1-1} \left\{ \begin{array}{c} \text{critical points of} \\ \psi^{(N)} : X \to \mathbb{R}, \ \psi^{(N)}(x) = f^{(N)}(x, x) \end{array} \right\}$$

Knowing that φ is generated by f, does $\varphi^{(2)}$ have a generating function? The answer is a partial yes:

Fix $x, y \in X$. Define a map

$$X \longrightarrow \mathbb{R}$$
$$z \longmapsto f(x,z) + f(z,y) .$$

Suppose that this map has a unique critical point z_0, and that z_0 is nondegenerate. Let

$$f^{(2)}(x,y) := f(x,z_0) + f(z_0,y) .$$

Proposition 5.2. *The function $f^{(2)} : X \times X \to \mathbb{R}$ is smooth and is a generating function for $\varphi^{(2)}$ if we assume that, for each $\xi \in T_x^* X$, there is a unique $y \in X$ for which $d_x f^{(2)} = \xi$.*

Proof. The point z_0 is given implicitly by $d_y f(x,z_0) + d_x f(z_0,y) = 0$. The nondegeneracy condition is

$$\det \left[\frac{\partial}{\partial z_i} \left(\frac{\partial f}{\partial y_j}(x,z) + \frac{\partial f}{\partial x_j}(z,y) \right) \right] \neq 0 .$$

By the implicit function theorem, $z_0 = z_0(x,y)$ is smooth.

As for the second assertion, $f^{(2)}(x,y)$ is a generating function for $\varphi^{(2)}$ if and only if

$$\varphi^{(2)}(x, d_x f^{(2)}) = (y, -d_y f^{(2)})$$

(assuming that, for each $\xi \in T_x^* X$, there is a unique $y \in X$ for which $d_x f^{(2)} = \xi$). Since φ is generated by f, and z_0 is critical, we obtain

$$\varphi^{(2)}(x, d_x f^{(2)}(x,y)) = \varphi(\underbrace{\varphi(x, d_x f^{(2)}(x,y))}_{=d_x f(x,z_0)}) = \varphi(z_0, -d_y f(x,z_0))$$

$$= \varphi(z_0, d_x f(z_0,y)) \quad = (y, \underbrace{-d_y f(z_0,y)}_{=-d_y f^{(2)}(x,y)}) .$$

$$\square$$

Exercise. What is a generating function for $\varphi^{(3)}$?
   Hint: Suppose that the function

$$X \times X \longrightarrow \mathbb{R}$$
$$(z,u) \longmapsto f(x,z) + f(z,u) + f(u,y)$$

has a unique critical point (z_0, u_0), and that it is a nondegenerate critical point. Let $f^{(3)}(x,y) = f(x,z_0) + f(z_0,u_0) + f(u_0,y)$.   ◇

5.2 Billiards

Let $\chi : \mathbb{R} \to \mathbb{R}^2$ be a smooth plane curve which is 1-periodic, i.e., $\chi(s+1) = \chi(s)$, and parametrized by arc-length, i.e., $\left|\frac{d\chi}{ds}\right| = 1$. Assume that the region Y enclosed by χ is *convex*, i.e., for any $s \in \mathbb{R}$, the tangent line $\{\chi(s) + t\frac{d\chi}{ds} \mid t \in \mathbb{R}\}$ intersects $X := \partial Y$ (= the image of χ) at only the point $\chi(s)$.

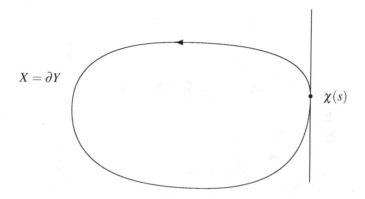

Suppose that we throw a ball into Y rolling with constant velocity and bouncing off the boundary with the usual law of reflection. This determines a map

$$\varphi : \mathbb{R}/\mathbb{Z} \times (-1,1) \longrightarrow \mathbb{R}/\mathbb{Z} \times (-1,1)$$
$$(x,v) \longmapsto (y,w)$$

by the rule

when the ball bounces off $\chi(x)$ with angle $\theta = \arccos v$, it will next collide with $\chi(y)$ and bounce off with angle $v = \arccos w$.

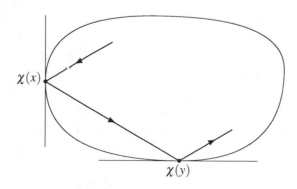

Let $f : \mathbb{R}/\mathbb{Z} \times \mathbb{R}/\mathbb{Z} \to \mathbb{R}$ be defined by $f(x,y) = -|\chi(x) - \chi(y)|$; f is smooth off the diagonal. Use χ to identify $\mathbb{R}/\mathbb{Z}$ with the image curve X.

Suppose that $\varphi(x,v) = (y,w)$, i.e., (x,v) and (y,w) are successive points on the orbit described by the ball. Then

$$
\begin{cases}
\dfrac{df}{dx} = -\dfrac{x-y}{|x-y|} \text{ projected onto } T_x X = \quad v \\[3mm]
\dfrac{df}{dy} = -\dfrac{y-x}{|x-y|} \text{ projected onto } T_y X = -w
\end{cases}
$$

or, equivalently,

$$
\begin{cases}
\dfrac{d}{ds} f(\chi(s),y) = \dfrac{y-x}{|x-y|} \cdot \dfrac{d\chi}{ds} = \quad \cos\theta = \quad v \\[3mm]
\dfrac{d}{ds} f(x,\chi(s)) = \dfrac{x-y}{|x-y|} \cdot \dfrac{d\chi}{ds} = -\cos\nu = -w .
\end{cases}
$$

We conclude that f is a generating function for φ. Similar approaches work for higher dimensional billiards problems.

Periodic points are obtained by finding critical points of

$$\underbrace{X \times \ldots \times X}_{N} \longrightarrow \mathbb{R} , \qquad\qquad N > 1$$

$$(x_1, \ldots, x_N) \longmapsto f(x_1,x_2) + f(x_2,x_3) + \ldots + f(x_{N-1},x_N) + f(x_N,x_1)$$
$$= |x_1 - x_2| + \ldots + |x_{N-1} - x_N| + |x_N - x_1| ,$$

that is, by finding the N-sided (generalized) polygons inscribed in X of critical perimeter.

Notice that

$$\mathbb{R}/\mathbb{Z} \times (-1,1) \simeq \{(x,v) \mid x \in X, v \in T_x X, |v| < 1\} \simeq A$$

is the open unit tangent ball bundle of a circle X, that is, an open annulus A. The map $\varphi : A \to A$ is area-preserving.

5.3 Poincaré Recurrence

Theorem 5.3. (Poincaré Recurrence Theorem)   *Suppose that $\varphi : A \to A$ is an area-preserving diffeomorphism of a finite-area manifold A. Let $p \in A$, and let $\mathcal{U}$ be a neighborhood of p. Then there is $q \in \mathcal{U}$ and a positive integer N such that $\varphi^{(N)}(q) \in \mathcal{U}$.*

Proof. Let $\mathcal{U}_0 = \mathcal{U}, \mathcal{U}_1 = \varphi(\mathcal{U}), \mathcal{U}_2 = \varphi^{(2)}(\mathcal{U}), \ldots$. If all of these sets were disjoint, then, since Area $(\mathcal{U}_i) =$ Area $(\mathcal{U}) > 0$ for all i, we would have

$$\text{Area } A \geq \text{ Area } (\mathcal{U}_0 \cup \mathcal{U}_1 \cup \mathcal{U}_2 \cup \ldots) = \sum_i \text{ Area } (\mathcal{U}_i) = \infty .$$

To avoid this contradiction we must have $\varphi^{(k)}(\mathcal{U}) \cap \varphi^{(l)}(\mathcal{U}) \neq \emptyset$ for some $k > l$, which implies $\varphi^{(k-l)}(\mathcal{U}) \cap \mathcal{U} \neq \emptyset$. □

Hence, eternal return applies to billiards...

Remark. Theorem 5.3 clearly generalizes to volume-preserving diffeomorphisms in higher dimensions. ◇

Theorem 5.4. (Poincaré's Last Geometric Theorem) *Suppose $\varphi : A \to A$ is an area-preserving diffeomorphism of the closed annulus $A = \mathbb{R}/\mathbb{Z} \times [-1, 1]$ which preserves the two components of the boundary, and twists them in opposite directions. Then φ has at least two fixed points.*

This theorem was proved in 1913 by Birkhoff, and hence is also called the **Poincaré-Birkhoff theorem**. It has important applications to dynamical systems and celestial mechanics. The Arnold conjecture (1966) on the existence of fixed points for symplectomorphisms of compact manifolds (see Lecture 9) may be regarded as a generalization of the Poincaré-Birkhoff theorem. This conjecture has motivated a significant amount of recent research involving a more general notion of generating function; see, for instance, [34, 45].

Part III
Local Forms

Inspired by the elementary normal form in symplectic linear algebra (Theorem 1.1), we will go on to describe normal neighborhoods of a point (the Darboux theorem) and of a lagrangian submanifold (the Weinstein theorems), inside a symplectic manifold. The main tool is the Moser trick, explained in Lecture 7, which leads to the crucial Moser theorems and which is at the heart of many arguments in symplectic geometry.

In order to prove the normal forms, we need the (non-symplectic) ingredients discussed in Lecture 6; for more on these topics, see, for instance, [18, 55, 96].

Chapter 6
Preparation for the Local Theory

6.1 Isotopies and Vector Fields

Let M be a manifold, and $\rho : M \times \mathbb{R} \to M$ a map, where we set $\rho_t(p) := \rho(p,t)$.

Definition 6.1. The map ρ is an *isotopy* if each $\rho_t : M \to M$ is a diffeomorphism, and $\rho_0 = \mathrm{id}_M$.

Given an isotopy ρ, we obtain a **time-dependent vector field**, that is, a family of vector fields $v_t, t \in \mathbb{R}$, which at $p \in M$ satisfy

$$v_t(p) = \frac{d}{ds}\rho_s(q)\Big|_{s=t} \qquad \text{where} \qquad q = \rho_t^{-1}(p) \, ,$$

i.e.,

$$\frac{d\rho_t}{dt} = v_t \circ \rho_t \, .$$

Conversely, given a time-dependent vector field v_t, if M is compact or if the v_t's are compactly supported, there exists an isotopy ρ satisfying the previous ordinary differential equation.

Suppose that M is compact. Then we have a one-to-one correspondence

$$\{\text{isotopies of } M\} \xrightarrow{1-1} \{\text{time-dependent vector fields on } M\}$$
$$\rho_t, \, t \in \mathbb{R} \longleftrightarrow v_t, \, t \in \mathbb{R}$$

Definition 6.2. When $v_t = v$ is independent of t, the associated isotopy is called the *exponential map* or the *flow* of v and is denoted $\exp tv$; i.e., $\{\exp tv : M \to M \mid t \in \mathbb{R}\}$ is the unique smooth family of diffeomorphisms satisfying

$$\exp tv|_{t=0} = \mathrm{id}_M \quad \text{and} \quad \frac{d}{dt}(\exp tv)(p) = v(\exp tv(p)) \, .$$

Definition 6.3. The *Lie derivative* is the operator

$$\mathcal{L}_v : \Omega^k(M) \longrightarrow \Omega^k(M) \qquad \text{defined by} \qquad \mathcal{L}_v\omega := \frac{d}{dt}(\exp tv)^*\omega|_{t=0} \, .$$

When a vector field v_t is time-dependent, its flow, that is, the corresponding isotopy ρ, still locally exists by Picard's theorem. More precisely, in the neighborhood of any point p and for sufficiently small time t, there is a one-parameter family of local diffeomorphisms ρ_t satisfying

$$\frac{d\rho_t}{dt} = v_t \circ \rho_t \qquad \text{and} \qquad \rho_0 = \text{id} \, .$$

Hence, we say that the **Lie derivative** by v_t is

$$\mathcal{L}_{v_t} : \Omega^k(M) \longrightarrow \Omega^k(M) \qquad \text{defined by} \qquad \mathcal{L}_{v_t}\omega := \frac{d}{dt}(\rho_t)^*\omega|_{t=0} \, .$$

Exercise. Prove the **Cartan magic formula**,

$$\mathcal{L}_v\omega = \iota_v d\omega + d\iota_v\omega \, ,$$

and the formula

$$\frac{d}{dt}\rho_t^*\omega = \rho_t^*\mathcal{L}_{v_t}\omega \, , \qquad\qquad (\star)$$

where ρ is the (local) isotopy generated by v_t. A good strategy for each formula is to follow the steps:

1. Check the formula for 0-forms $\omega \in \Omega^0(M) = C^\infty(M)$.
2. Check that both sides commute with d.
3. Check that both sides are derivations of the algebra $(\Omega^*(M), \wedge)$. For instance, check that

$$\mathcal{L}_v(\omega \wedge \alpha) = (\mathcal{L}_v\omega) \wedge \alpha + \omega \wedge (\mathcal{L}_v\alpha) \, .$$

4. Notice that, if $\mathcal{U}$ is the domain of a coordinate system, then $\Omega^\bullet(\mathcal{U})$ is generated as an algebra by $\Omega^0(\mathcal{U})$ and $d\Omega^0(\mathcal{U})$, i.e., every element in $\Omega^\bullet(\mathcal{U})$ is a linear combination of wedge products of elements in $\Omega^0(\mathcal{U})$ and elements in $d\Omega^0(\mathcal{U})$.

We will need the following improved version of formula $(\star)$.

Proposition 6.4. *For a smooth family ω_t, $t \in \mathbb{R}$, of d-forms, we have*

$$\frac{d}{dt}\rho_t^*\omega_t = \rho_t^*\left(\mathcal{L}_{v_t}\omega_t + \frac{d\omega_t}{dt}\right) \, .$$

Proof. If $f(x,y)$ is a real function of two variables, by the chain rule we have

$$\frac{d}{dt} f(t,t) = \frac{d}{dx} f(x,t) \Big|_{x=t} + \frac{d}{dy} f(t,y) \Big|_{y=t} .$$

Therefore,

$$\frac{d}{dt} \rho_t^* \omega_t = \underbrace{\frac{d}{dx} \rho_x^* \omega_t \Big|_{x=t}}_{\rho_x^* \mathcal{L}_{v_x} \omega_t \big|_{x=t} \text{ by } (\star)} + \underbrace{\frac{d}{dy} \rho_t^* \omega_y \Big|_{y=t}}_{\rho_t^* \frac{d\omega_y}{dy} \big|_{y=t}}$$

$$= \rho_t^* \left(\mathcal{L}_{v_t} \omega_t + \frac{d\omega_t}{dt} \right) .$$

$\square$

6.2 Tubular Neighborhood Theorem

Let M be an n-dimensional manifold, and let X be a k-dimensional submanifold where $k < n$ and with inclusion map

$$i : X \hookrightarrow M .$$

At each $x \in X$, the tangent space to X is viewed as a subspace of the tangent space to M via the linear inclusion $di_x : T_x X \hookrightarrow T_x M$, where we denote $x = i(x)$. The quotient $N_x X := T_x M / T_x X$ is an $(n-k)$-dimensional vector space, known as the **normal space** to X at x. The **normal bundle** of X is

$$NX = \{ (x,v) \mid x \in X , v \in N_x X \} .$$

The set NX has the structure of a vector bundle over X of rank $n-k$ under the natural projection, hence as a manifold NX is n-dimensional. The zero section of NX,

$$i_0 : X \hookrightarrow NX , \qquad x \mapsto (x,0) ,$$

embeds X as a closed submanifold of NX. A neighborhood $\mathcal{U}_0$ of the zero section X in NX is called **convex** if the intersection $\mathcal{U}_0 \cap N_x X$ with each fiber is convex.

Theorem 6.5. *(Tubular Neighborhood Theorem) There exist a convex neighborhood $\mathcal{U}_0$ of X in NX, a neighborhood $\mathcal{U}$ of X in M, and a diffeomorphism $\varphi : \mathcal{U}_0 \to \mathcal{U}$ such that*

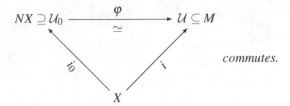

$$NX \supseteq \mathcal{U}_0 \xrightarrow[\simeq]{\varphi} \mathcal{U} \subseteq M$$

commutes.

Outline of the proof.

- *Case of $M = \mathbb{R}^n$, and X is a compact submanifold of $\mathbb{R}^n$.*

   Theorem 6.6. (ε-Neighborhood Theorem)
   Let $\mathcal{U}^\varepsilon = \{p \in \mathbb{R}^n : |p - q| < \varepsilon$ for some $q \in X\}$ be the set of points at a distance less than ε from X. Then, for ε sufficiently small, each $p \in \mathcal{U}^\varepsilon$ has a unique nearest point $q \in X$ (i.e., a unique $q \in X$ minimizing $|q - x|$).
   Moreover, setting $q = \pi(p)$, the map $\mathcal{U}^\varepsilon \xrightarrow{\pi} X$ is a (smooth) submersion with the property that, for all $p \in \mathcal{U}^\varepsilon$, the line segment $(1 - t)p + tq$, $0 \le t \le 1$, is in $\mathcal{U}^\varepsilon$.

   The proof is part of Homework 5. Here are some hints.
   At any $x \in X$, the *normal* space $N_x X$ may be regarded as an $(n - k)$-dimensional subspace of $\mathbb{R}^n$, namely the orthogonal complement in $\mathbb{R}^n$ of the tangent space to X at x:
   $$N_x X \simeq \{v \in \mathbb{R}^n : v \perp w, \text{ for all } w \in T_x X\}.$$

   We define the following open neighborhood of X in NX:

   $$NX^\varepsilon = \{(x, v) \in NX : |v| < \varepsilon\}.$$

Let
$$\begin{aligned} \exp: \quad NX &\longrightarrow \mathbb{R}^n \\ (x, v) &\longmapsto x + v. \end{aligned}$$

Restricted to the zero section, $\exp$ is the identity map on X.
Prove that, for ε sufficiently small, $\exp$ maps NX^ε diffeomorphically onto $\mathcal{U}^\varepsilon$, and show also that the diagram

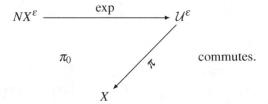

$$NX^\varepsilon \xrightarrow{\exp} \mathcal{U}^\varepsilon$$

π_0 $\qquad$ π $\qquad$ commutes.

$$X$$

- *Case where X is a compact submanifold of an arbitrary manifold M.*
   Put a riemannian metric g on M, and let $d(p, q)$ be the riemannian distance between $p, q \in M$. The ε-neighborhood of a compact submanifold X is

   $$\mathcal{U}^\varepsilon = \{p \in M \mid d(p, q) < \varepsilon \text{ for some } q \in X\}.$$

Prove the ε-neighborhood theorem in this setting: for ε small enough, the following assertions hold.

– Any $p \in \mathcal{U}^\varepsilon$ has a unique point $q \in X$ with minimal $d(p,q)$. Set $q = \pi(p)$.

– The map $\mathcal{U}^\varepsilon \xrightarrow{\pi} X$ is a submersion and, for all $p \in \mathcal{U}^\varepsilon$, there is a unique geodesic curve γ joining p to $q = \pi(p)$.

– The normal space to X at $x \in X$ is naturally identified with a subspace of $T_x M$:

$$N_x X \simeq \{ v \in T_x M \mid g_x(v,w) = 0 \text{ , for any } w \in T_x X \} \ .$$

Let $NX^\varepsilon = \{ (x,v) \in NX \mid \sqrt{g_x(v,v)} < \varepsilon \}$.

– Define $\exp : NX^\varepsilon \to M$ by $\exp(x,v) = \gamma(1)$, where $\gamma : [0,1] \to M$ is the geodesic with $\gamma(0) = x$ and $\frac{d\gamma}{dt}(0) = v$. Then $\exp$ maps NX^ε diffeomorphically to $\mathcal{U}^\varepsilon$.

• *General case.*

When X is not compact, adapt the previous argument by replacing ε by an appropriate continuous function $\varepsilon : X \to \mathbb{R}^+$ which tends to zero fast enough as x tends to infinity.

$\square$

Restricting to the subset $\mathcal{U}^0 \subseteq NX$ from the tubular neighborhood theorem, we obtain a submersion $\mathcal{U}_0 \xrightarrow{\pi_0} X$ with all fibers $\pi_0^{-1}(x)$ convex. We can carry this fibration to $\mathcal{U}$ by setting $\pi = \pi_0 \circ \varphi^{-1}$:

$$
\begin{array}{ccc}
\mathcal{U}_0 \ \subseteq NX \ \text{ is a fibration} & \Longrightarrow & \mathcal{U} \ \subseteq M \ \text{ is a fibration} \\
\pi_0 \downarrow & & \pi \downarrow \\
X & & X
\end{array}
$$

This is called the **tubular neighborhood fibration**.

6.3 Homotopy Formula

Let $\mathcal{U}$ be a tubular neighborhood of a submanifold X in M. The restriction $i^* : H^d_{\text{deRham}}(\mathcal{U}) \to H^d_{\text{deRham}}(X)$ by the inclusion map is surjective. As a corollary of the tubular neighborhood fibration, i^* is also injective: this follows from the homotopy-invariance of de Rham cohomology.

Corollary 6.7. *For any degree ℓ, $H^\ell_{\text{deRham}}(\mathcal{U}) \simeq H^\ell_{\text{deRham}}(X)$.*

At the level of forms, this means that, if ω is a closed ℓ-form on $\mathcal{U}$ and $i^*\omega$ is exact on X, then ω is exact. We will need the following related result.

Proposition 6.8. *If a closed ℓ-form ω on $\mathcal{U}$ has restriction $i^*\omega = 0$, then ω is exact, i.e., $\omega = d\mu$ for some $\mu \in \Omega^{d-1}(\mathcal{U})$. Moreover, we can choose μ such that $\mu_x = 0$ at all $x \in X$.*

Proof. Via $\varphi : \mathcal{U}_0 \xrightarrow{\simeq} \mathcal{U}$, it is equivalent to work over $\mathcal{U}_0$. Define for every $0 \leq t \leq 1$ a map

$$\rho_t : \quad \mathcal{U}_0 \longrightarrow \mathcal{U}_0$$
$$(x, v) \longmapsto (x, tv) .$$

This is well-defined since $\mathcal{U}_0$ is convex. The map ρ_1 is the identity, $\rho_0 = i_0 \circ \pi_0$, and each ρ_t fixes X, that is, $\rho_t \circ i_0 = i_0$. We hence say that the family $\{\rho_t \mid 0 \leq t \leq 1\}$ is a **homotopy** from $i_0 \circ \pi_0$ to the identity fixing X. The map $\pi_0 : \mathcal{U}_0 \to X$ is called a **retraction** because $\pi_0 \circ i_0$ is the identity. The submanifold X is then called a **deformation retract** of $\mathcal{U}$.

A (de Rham) **homotopy operator** between $\rho_0 = i_0 \circ \pi_0$ and $\rho_1 = \mathrm{id}$ is a linear map

$$Q : \Omega^d(\mathcal{U}_0) \longrightarrow \Omega^{d-1}(\mathcal{U}_0)$$

satisfying the **homotopy formula**

$$\mathrm{Id} - (i_0 \circ \pi_0)^* = dQ + Qd .$$

When $d\omega = 0$ and $i_0^*\omega = 0$, the operator Q gives $\omega = dQ\omega$, so that we can take $\mu = Q\omega$. A concrete operator Q is given by the formula:

$$Q\omega = \int_0^1 \rho_t^*(\iota_{v_t}\omega) \, dt ,$$

where v_t, at the point $q = \rho_t(p)$, is the vector tangent to the curve $\rho_s(p)$ at $s = t$. The proof that Q satisfies the homotopy formula is below.

In our case, for $x \in X$, $\rho_t(x) = x$ (all t) is the constant curve, so v_t vanishes at all x for all t, hence $\mu_x = 0$. $\qquad\qquad\square$

To check that Q above satisfies the homotopy formula, we compute

$$Qd\omega + dQ\omega = \int_0^1 \rho_t^*(\iota_{v_t}d\omega)dt + d\int_0^1 \rho_t^*(\iota_{v_t}\omega)dt$$

$$= \int_0^1 \rho_t^*(\underbrace{\iota_{v_t}d\omega + d\iota_{v_t}\omega}_{\mathcal{L}_{v_t}\omega})dt ,$$

where $\mathcal{L}_v$ denotes the Lie derivative along v (reviewed in the next section), and we used the Cartan magic formula: $\mathcal{L}_v\omega = \iota_v d\omega + d\iota_v\omega$. The result now follows from

$$\frac{d}{dt}\rho_t^*\omega = \rho_t^*\mathcal{L}_{v_t}\omega$$

and from the fundamental theorem of calculus:

$$Qd\omega + dQ\omega = \int_0^1 \frac{d}{dt}\rho_t^*\omega \, dt = \rho_1^*\omega - \rho_0^*\omega .$$

Homework 5: Tubular Neighborhoods in $\mathbb{R}^n$

1. Let X be a k-dimensional submanifold of an n-dimensional manifold M. Let x be a point in X. The **normal space** to X at x is the quotient space

$$N_xX = T_xM/T_xX\ ,$$

and the **normal bundle** of X in M is the vector bundle NX over X whose fiber at x is N_xX.

   (a) Prove that NX is indeed a vector bundle.
   (b) If M is $\mathbb{R}^n$, show that N_xX can be identified with the usual "normal space" to X in $\mathbb{R}^n$, that is, the orthogonal complement in $\mathbb{R}^n$ of the tangent space to X at x.

2. Let X be a k-dimensional compact submanifold of $\mathbb{R}^n$. Prove the **tubular neighborhood theorem** in the following form.

   (a) Given $\varepsilon > 0$ let $\mathcal{U}_\varepsilon$ be the set of all points in $\mathbb{R}^n$ which are at a distance less than ε from X. Show that, for ε sufficiently small, every point $p \in \mathcal{U}_\varepsilon$ has a *unique* nearest point $\pi(p) \in X$.
   (b) Let $\pi : \mathcal{U}_\varepsilon \to X$ be the map defined in (a) for ε sufficiently small. Show that, if $p \in \mathcal{U}_\varepsilon$, then the line segment $(1-t) \cdot p + t \cdot \pi(p)$, $0 \le t \le 1$, joining p to $\pi(p)$ lies in $\mathcal{U}_\varepsilon$.
   (c) Let $NX_\varepsilon = \{(x,v) \in NX \text{ such that } |v| < \varepsilon\}$. Let $\exp : NX \to \mathbb{R}^n$ be the map $(x,v) \mapsto x+v$, and let $\mathrm{v} : NX_\varepsilon \to X$ be the map $(x,v) \mapsto x$. Show that, for ε sufficiently small, $\exp$ maps NX_ε diffeomorphically onto $\mathcal{U}_\varepsilon$, and show also that the following diagram commutes:

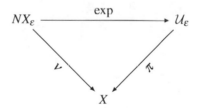

3. Suppose that the manifold X in the previous exercise is not compact. Prove that the assertion about exp is still true provided we replace ε by a continuous function

$$\varepsilon : X \to \mathbb{R}^+$$

which tends to zero fast enough as x tends to infinity.

Chapter 7
Moser Theorems

7.1 Notions of Equivalence for Symplectic Structures

Let M be a $2n$-dimensional manifold with two symplectic forms ω_0 and ω_1, so that (M, ω_0) and (M, ω_1) are two symplectic manifolds.

Definition 7.1. We say that

- (M, ω_0) and (M, ω_1) are **symplectomorphic** if there is a diffeomorphism $\varphi : M \to M$ with $\varphi^* \omega_1 = \omega_0$;
- (M, ω_0) and (M, ω_1) are **strongly isotopic** if there is an isotopy $\rho_t : M \to M$ such that $\rho_1^* \omega_1 = \omega_0$;
- (M, ω_0) and (M, ω_1) are **deformation-equivalent** if there is a smooth family ω_t of symplectic forms joining ω_0 to ω_1,
- (M, ω_0) and (M, ω_1) are **isotopic** if they are deformation-equivalent with $[\omega_t]$ independent of t.

Clearly, we have

$$\text{strongly isotopic} \implies \text{symplectomorphic}, \quad \text{and}$$

$$\text{isotopic} \implies \text{deformation-equivalent}.$$

We also have

$$\text{strongly isotopic} \implies \text{isotopic}$$

because, if $\rho_t : M \to M$ is an isotopy such that $\rho_1^* \omega_1 = \omega_0$, then the set $\omega_t := \rho_t^* \omega_1$ is a smooth family of symplectic forms joining ω_1 to ω_0 and $[\omega_t] = [\omega_1]$, $\forall t$, by the homotopy invariance of de Rham cohomology. As we will see below, the Moser theorem states that, on a compact manifold,

$$\text{isotopic} \implies \text{strongly isotopic}.$$

7.2 Moser Trick

Problem. Given a $2n$-dimensional manifold M, a k-dimensional submanifold X, neighborhoods $\mathcal{U}_0, \mathcal{U}_1$ of X, and symplectic forms ω_0, ω_1 on $\mathcal{U}_0, \mathcal{U}_1$, does there exist a symplectomorphism preserving X? More precisely, does there exist a diffeomorphism $\varphi : \mathcal{U}_0 \to \mathcal{U}_1$ with $\varphi^* \omega_1 = \omega_0$ and $\varphi(X) = X$?

At the two extremes, we have:
Case $X = point$:   Darboux theorem – see Lecture 8.
Case $X = M$:   Moser theorem – discussed here:

Let M be a *compact* manifold with symplectic forms ω_0 and ω_1.

– Are (M, ω_0) and (M, ω_1) symplectomorphic?
I.e., does there exist a diffeomorphism $\varphi : M \to M$ such that $\varphi_1^* \omega_0 = \omega_1$?

Moser asked whether we can find such an φ which is homotopic to id_M. A necessary condition is $[\omega_0] = [\omega_1] \in H^2(M; \mathbb{R})$ because: if $\varphi \sim \mathrm{id}_M$, then, by the homotopy formula, there exists a homotopy operator Q such that

$$\mathrm{id}_M^* \omega_1 - \varphi^* \omega_1 = dQ\omega_1 + Q\underbrace{d\omega_1}_{0}$$

$$\implies \quad \omega_1 = \varphi^* \omega_1 + d(Q\omega_1)$$

$$\implies \quad [\omega_1] = [\varphi^* \omega_1] = [\omega_0] .$$

– If $[\omega_0] = [\omega_1]$, does there exist a diffeomorphism φ homotopic to id_M such that $\varphi^* \omega_1 = \omega_0$?

Moser [87] proved that the answer is yes, with a further hypothesis as in Theorem 7.2. McDuff showed that, in general, the answer is no; for a counterexample, see Example 7.23 in [83].

Theorem 7.2. (Moser Theorem – Version I) *Suppose that M is compact, $[\omega_0] = [\omega_1]$ and that the 2-form $\omega_t = (1-t)\omega_0 + t\omega_1$ is symplectic for each $t \in [0,1]$. Then there exists an isotopy $\rho : M \times \mathbb{R} \to M$ such that $\rho_t^* \omega_t = \omega_0$ for all $t \in [0,1]$.*

In particular, $\varphi = \rho_1 : M \xrightarrow{\simeq} M$, satisfies $\varphi^* \omega_1 = \omega_0$.

The following argument, due to Moser, is extremely useful; it is known as the **Moser trick**.

Proof. Suppose that there exists an isotopy $\rho : M \times \mathbb{R} \to M$ such that $\rho_t^* \omega_t = \omega_0$, $0 \le t \le 1$. Let

$$v_t = \frac{d\rho_t}{dt} \circ \rho_t^{-1} , \qquad t \in \mathbb{R} .$$

Then

$$0 = \frac{d}{dt}(\rho_t^* \omega_t) = \rho_t^* \left(\mathcal{L}_{v_t} \omega_t + \frac{d\omega_t}{dt} \right)$$

$$\Longleftrightarrow \qquad \mathcal{L}_{v_t} \omega_t + \frac{d\omega_t}{dt} = 0 . \qquad (\star)$$

Suppose conversely that we can find a smooth time-dependent vector field v_t, $t \in \mathbb{R}$, such that $(\star)$ holds for $0 \le t \le 1$. Since M is compact, we can integrate v_t to an isotopy $\rho : M \times \mathbb{R} \to M$ with

$$\frac{d}{dt}(\rho_t^* \omega_t) = 0 \quad \Longrightarrow \quad \rho_t^* \omega_t = \rho_0^* \omega_0 = \omega_0 .$$

So everything boils down to solving $(\star)$ for v_t.

First, from $\omega_t = (1-t)\omega_0 + t\omega_1$, we conclude that

$$\frac{d\omega_t}{dt} = \omega_1 - \omega_0 .$$

Second, since $[\omega_0] = [\omega_1]$, there exists a 1-form μ such that

$$\omega_1 - \omega_0 = d\mu .$$

Third, by the Cartan magic formula, we have

$$\mathcal{L}_{v_t} \omega_t = d\iota_{v_t} \omega_t + \underbrace{\iota_{v_t} d\omega_t}_{0} .$$

Putting everything together, we must find v_t such that

$$d\iota_{v_t} \omega_t + d\mu = 0 .$$

It is sufficient to solve $\iota_{v_t} \omega_t + \mu = 0$. By the nondegeneracy of ω_t, we can solve this pointwise, to obtain a unique (smooth) v_t. $\qquad\qquad\qquad\qquad\qquad\qquad$ $\square$

Theorem 7.3. (Moser Theorem – Version II) *Let M be a compact manifold with symplectic forms ω_0 and ω_1. Suppose that ω_t, $0 \le t \le 1$, is a smooth family of closed 2-forms joining ω_0 to ω_1 and satisfying:*

(1) cohomology assumption: $[\omega_t]$ is independent of t, i.e., $\frac{d}{dt}[\omega_t] = \left[\frac{d}{dt}\omega_t\right] = 0$,
(2) nondegeneracy assumption: ω_t is nondegenerate for $0 \le t \le 1$.

Then there exists an isotopy $\rho : M \times \mathbb{R} \to M$ such that $\rho_t^* \omega_t = \omega_0$, $0 \le t \le 1$.

Proof. (Moser trick) We have the following implications from the hypotheses:

(1) $\Longrightarrow$ $\exists$ family of 1-forms μ_t such that

$$\frac{d\omega_t}{dt} = d\mu_t , \quad 0 \le t \le 1 .$$

We can indeed find a *smooth* family of 1-forms μ_t such that $\frac{d\omega_t}{dt} = d\mu_t$. The argument involves the Poincaré lemma for compactly-supported forms, together with the Mayer-Vietoris sequence in order to use induction on the number of charts in a good cover of M. For a sketch of the argument, see page 95 in [83].

(2) $\implies$ $\exists$ unique family of vector fields v_t such that

$$\iota_{v_t}\omega_t + \mu_t = 0 \qquad \textbf{(Moser equation)} .$$

Extend v_t to all $t \in \mathbb{R}$. Let ρ be the isotopy generated by v_t (ρ exists by compactness of M). Then we indeed have

$$\frac{d}{dt}(\rho_t^*\omega_t) = \rho_t^*(\mathcal{L}_{v_t}\omega_t + \frac{d\omega_t}{dt}) = \rho_t^*(d\iota_{v_t}\omega_t + d\mu_t) = 0 .$$

$\square$

The compactness of M was used to be able to integrate v_t for all $t \in \mathbb{R}$. If M is *not* compact, we need to check the existence of a solution ρ_t for the differential equation $\frac{d\rho_t}{dt} = v_t \circ \rho_t$ for $0 \leq t \leq 1$.

Picture. Fix $c \in H^2(M)$. Define $S_c = \{$symplectic forms ω in M with $[\omega] = c\}$. The Moser theorem implies that, on a compact manifold, all symplectic forms on the same path-connected component of S_c are symplectomorphic.

7.3 Moser Relative Theorem

Theorem 7.4. *(**Moser Theorem – Relative Version**)*   *Let M be a manifold, X a compact submanifold of M, $i : X \hookrightarrow M$ the inclusion map, ω_0 and ω_1 symplectic forms in M.*

Hypothesis: $\omega_0|_p = \omega_1|_p$, $\forall p \in X$.
Conclusion: *There exist neighborhoods $\mathcal{U}_0,\mathcal{U}_1$ of X in M ,*
and a diffeomorphism $\varphi : \mathcal{U}_0 \to \mathcal{U}_1$ such that

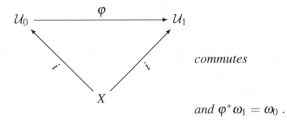

and $\varphi^*\omega_1 = \omega_0$.

Proof.

1. Pick a tubular neighborhood $\mathcal{U}_0$ of X. The 2-form $\omega_1 - \omega_0$ is closed on $\mathcal{U}_0$, and $(\omega_1 - \omega_0)_p = 0$ at all $p \in X$. By the homotopy formula on the tubular neighborhood, there exists a 1-form μ on $\mathcal{U}_0$ such that $\omega_1 - \omega_0 = d\mu$ and $\mu_p = 0$ at all $p \in X$.

2. Consider the family $\omega_t = (1-t)\omega_0 + t\omega_1 = \omega_0 + td\mu$ of closed 2-forms on $\mathcal{U}_0$. Shrinking $\mathcal{U}_0$ if necessary, we can assume that ω_t is symplectic for $0 \leq t \leq 1$.

3. Solve the Moser equation: $\iota_{v_t}\omega_t = -\mu$. Notice that $v_t = 0$ on X.

4. Integrate v_t. Shrinking $\mathcal{U}_0$ again if necessary, there exists an isotopy $\rho : \mathcal{U}_0 \times [0,1] \to M$ with $\rho_t^*\omega_t = \omega_0$, for all $t \in [0,1]$. Since $v_t|_X = 0$, we have $\rho_t|_X = \mathrm{id}_X$.

   Set $\varphi = \rho_1, \mathcal{U}_1 = \rho_1(\mathcal{U}_0)$. $\qquad\qquad\qquad\qquad\qquad\qquad\qquad\square$

Exercise. Prove the Darboux theorem. (Hint: apply the relative version of the Moser theorem to $X = \{p\}$, as in the next lecture.) $\qquad\qquad\qquad\qquad\Diamond$

Chapter 8
Darboux-Moser-Weinstein Theory

8.1 Darboux Theorem

Theorem 8.1. (Darboux) *Let (M, ω) be a symplectic manifold, and let p be any point in M. Then we can find a coordinate system $(\mathcal{U}, x_1, \ldots, x_n, y_1, \ldots y_n)$ centered at p such that on $\mathcal{U}$*

$$\omega = \sum_{i=1}^{n} dx_i \wedge dy_i .$$

As a consequence of Theorem 8.1, if we prove for $(\mathbb{R}^{2n}, \sum dx_i \wedge dy_i)$ a local assertion which is invariant under symplectomorphisms, then that assertion holds for any symplectic manifold.

Proof. Apply the Moser relative theorem (Theorem 7.4) to $X = \{p\}$:

Use any symplectic basis for $T_p M$ to construct coordinates $(x'_1, \ldots, x'_n, y'_1, \ldots y'_n)$ centered at p and valid on some neighborhood $\mathcal{U}'$, so that

$$\omega_p = \sum dx'_i \wedge dy'_i \big|_p .$$

There are two symplectic forms on $\mathcal{U}'$: the given $\omega_0 = \omega$ and $\omega_1 = \sum dx'_i \wedge dy'_i$. By the Moser theorem, there are neighborhoods $\mathcal{U}_0$ and $\mathcal{U}_1$ of p, and a diffeomorphism $\varphi : \mathcal{U}_0 \to \mathcal{U}_1$ such that

$$\varphi(p) = p \quad \text{and} \quad \varphi^*(\sum dx'_i \wedge dy'_i) = \omega .$$

Since $\varphi^*(\sum dx'_i \wedge dy'_i) = \sum d(x'_i \circ \varphi) \wedge d(y'_i \circ \varphi)$, we only need to set new coordinates $x_i = x'_i \circ \varphi$ and $y_i = y'_i \circ \varphi$. $\qquad \square$

If in the Moser relative theorem (Theorem 7.4) we assume instead

> Hypothesis: X is an n-dimensional submanifold with
> $i^* \omega_0 = i^* \omega_1 = 0$ where $i : X \hookrightarrow M$ is inclusion, i.e.,
> X is a submanifold lagrangian for ω_0 and ω_1 ,

then Weinstein [104] proved that the conclusion still holds. We need some algebra for the Weinstein theorem.

8.2 Lagrangian Subspaces

Suppose that U, W are n-dimensional vector spaces, and $\Omega : U \times W \to \mathbb{R}$ is a bilinear pairing; the map Ω gives rise to a linear map $\widetilde{\Omega} : U \to W^*$, $\widetilde{\Omega}(u) = \Omega(u, \cdot)$. Then Ω is nondegenerate if and only if $\widetilde{\Omega}$ is bijective.

Proposition 8.2. *Suppose that (V, Ω) is a $2n$-dimensional symplectic vector space. Let U be a lagrangian subspace of (V, Ω) (i.e., $\Omega|_{U \times U} = 0$ and U is n-dimensional). Let W be any vector space complement to U, not necessarily lagrangian.*

Then from W we can canonically *build a* lagrangian *complement to U.*

Proof. The pairing Ω gives a nondegenerate pairing $U \times W \overset{\Omega'}{\to} \mathbb{R}$. Therefore, $\widetilde{\Omega}' : U \to W^*$ is bijective. We look for a lagrangian complement to U of the form

$$W' = \{w + Aw \mid w \in W\} ,$$

the map $A : W \to U$ being linear. For W' to be lagrangian we need

$$\forall \, w_1, w_2 \in W , \quad \Omega(w_1 + Aw_1, w_2 + Aw_2) = 0$$

$$\implies \Omega(w_1, w_2) + \Omega(w_1, Aw_2) + \Omega(Aw_1, w_2) + \underbrace{\Omega(\underbrace{Aw_1, Aw_2}_{\in U})}_{0} = 0$$

$$\implies \Omega(w_1, w_2) = \Omega(Aw_2, w_1) - \Omega(Aw_1, w_2)$$
$$= \widetilde{\Omega}'(Aw_2)(w_1) - \widetilde{\Omega}'(Aw_1)(w_2) .$$

Let $A' = \widetilde{\Omega}' \circ A : W \to W^*$, and look for A' such that

$$\forall \, w_1, w_2 \in W , \quad \Omega(w_1, w_2) = A'(w_2)(w_1) - A'(w_1)(w_2) .$$

The canonical choice is $A'(w) = -\frac{1}{2}\Omega(w, \cdot)$. Then set $A = (\widetilde{\Omega}')^{-1} \circ A'$. $\qquad \square$

Proposition 8.3. *Let V be a $2n$-dimensional vector space, let Ω_0 and Ω_1 be symplectic forms in V, let U be a subspace of V lagrangian for Ω_0 and Ω_1, and let W be any complement to U in V. Then from W we can canonically construct a linear isomorphism $L : V \overset{\cong}{\to} V$ such that $L|_U = \mathrm{Id}_U$ and $L^* \Omega_1 = \Omega_0$.*

Proof. From W we canonically obtain complements W_0 and W_1 to U in V such that W_0 is lagrangian for Ω_0 and W_1 is lagrangian for Ω_1. The nondegenerate bilinear pairings

$$W_0 \times U \overset{\Omega_0}{\longrightarrow} \mathbb{R} \qquad \text{give isomorphisms} \qquad \widetilde{\Omega}_0 : W_0 \overset{\cong}{\longrightarrow} U^*$$
$$W_1 \times U \overset{\Omega_1}{\longrightarrow} \mathbb{R} \qquad \qquad \qquad \qquad \widetilde{\Omega}_1 : W_1 \overset{\cong}{\longrightarrow} U^* .$$

Consider the diagram

$$
\begin{array}{ccc}
W_0 & \xrightarrow{\widetilde{\Omega}_0} & U^* \\
B \downarrow & & \downarrow \text{id} \\
W_1 & \xrightarrow{\widetilde{\Omega}_1} & U^*
\end{array}
$$

where the linear map B satisfies $\widetilde{\Omega}_1 \circ B = \widetilde{\Omega}_0$, i.e., $\Omega_0(w_0, u) = \Omega_1(Bw_0, u)$, $\forall w_0 \in W_0$, $\forall u \in U$. Extend B to the rest of V by setting it to be the identity on U:

$$
L := \text{Id}_U \oplus B : U \oplus W_0 \longrightarrow U \oplus W_1 .
$$

Finally, we check that $L^* \Omega_1 = \Omega_0$.

$$
\begin{aligned}
(L^* \Omega_1)(u \oplus w_0, u' \oplus w_0') &= \Omega_1(u \oplus Bw_0, u' \oplus Bw_0') \\
&= \Omega_1(u, Bw_0') + \Omega_1(Bw_0, u') \\
&= \Omega_0(u, w_0') + \Omega_0(w_0, u') \\
&= \Omega_0(u \oplus w_0, u' \oplus w_0') .
\end{aligned}
$$

$\square$

8.3 Weinstein Lagrangian Neighborhood Theorem

Theorem 8.4. *(Weinstein Lagrangian Neighborhood Theorem [104])* Let M be a $2n$-dimensional manifold, X a compact n-dimensional submanifold, $i : X \hookrightarrow M$ the inclusion map, and ω_0 and ω_1 symplectic forms on M such that $i^* \omega_0 = i^* \omega_1 = 0$, i.e., X is a lagrangian submanifold of both (M, ω_0) and (M, ω_1). Then there exist neighborhoods $\mathcal{U}_0$ and $\mathcal{U}_1$ of X in M and a diffeomorphism $\varphi : \mathcal{U}_0 \to \mathcal{U}_1$ such that

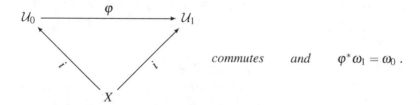

commutes   and $\varphi^* \omega_1 = \omega_0$.

The proof of the Weinstein theorem uses the Whitney extension theorem.

Theorem 8.5. *(Whitney Extension Theorem)* Let M be an n-dimensional manifold and X a k-dimensional submanifold with $k < n$. Suppose that at each $p \in X$ we are given a linear isomorphism $L_p : T_p M \xrightarrow{\cong} T_p M$ such that $L_p|_{T_p X} = \text{Id}_{T_p X}$ and L_p depends smoothly on p. Then there exists an embedding $h : \mathcal{N} \to M$ of some neighborhood $\mathcal{N}$ of X in M such that $h|_X = \text{id}_X$ and $dh_p = L_p$ for all $p \in X$.

The linear maps L serve as "germs" for the embedding.

Proof of the Weinstein theorem. Put a riemannian metric g on M; at each $p \in M$, $g_p(\cdot, \cdot)$ is a positive-definite inner product. Fix $p \in X$, and let $V = T_pM$, $U = T_pX$ and $W = U^\perp$ the orthocomplement of U in V relative to $g_p(\cdot, \cdot)$.

Since $i^*\omega_0 = i^*\omega_1 = 0$, the space U is a lagrangian subspace of both $(V, \omega_0|_p)$ and $(V, \omega_1|_p)$. By symplectic linear algebra, we canonically get from $U^\perp$ a linear isomorphism $L_p : T_pM \to T_pM$, such that $L_p|_{T_pX} = \mathrm{Id}_{T_pX}$ and $L_p^*\omega_1|_p = \omega_0|_p$. L_p varies smoothly with respect to p since our recipe is canonical!

By the Whitney theorem, there are a neighborhood $\mathcal{N}$ of X and an embedding $h : \mathcal{N} \hookrightarrow M$ with $h|_X = \mathrm{id}_X$ and $dh_p = L_p$ for $p \in X$. Hence, at any $p \in X$,

$$(h^*\omega_1)_p = (dh_p)^*\omega_1|_p = L_p^*\omega_1|_p = \omega_0|_p \ .$$

Applying the Moser relative theorem (Theorem 7.4) to ω_0 and $h^*\omega_1$, we find a neighborhood $\mathcal{U}_0$ of X and an embedding $f : \mathcal{U}_0 \to \mathcal{N}$ such that $f|_X = \mathrm{id}_X$ and $f^*(h^*\omega_1) = \omega_0$ on $\mathcal{U}_o$. Set $\varphi = h \circ f$. $\square$

Sketch of proof for the Whitney theorem.

Case $M = \mathbb{R}^n$:

For a compact k-dimensional submanifold X, take a neighborhood of the form

$$\mathcal{U}^\varepsilon = \{ p \in M \mid \text{distance } (p, X) \leq \varepsilon \} \ .$$

For ε sufficiently small so that any $p \in \mathcal{U}^\varepsilon$ has a unique nearest point in X, define a projection $\pi : \mathcal{U}^\varepsilon \to X$, $p \mapsto$ point on X closest to p. If $\pi(p) = q$, then $p = q + v$ for some $v \in N_qX$ where $N_qX = (T_qX)^\perp$ is the normal space at q; see Homework 5. Let

$$h : \mathcal{U}^\varepsilon \longrightarrow \mathbb{R}^n$$
$$p \longmapsto q + L_q v \ ,$$

where $q = \pi(p)$ and $v = p - \pi(p) \in N_qX$. Then $h_X = \mathrm{id}_X$ and $dh_p = L_p$ for $p \in X$. If X is not compact, replace ε by a continuous function $\varepsilon : X \to \mathbb{R}^+$ which tends to zero fast enough as x tends to infinity.

General case:

Choose a riemannian metric on M. Replace distance by riemannian distance, replace straight lines $q + tv$ by geodesics $\exp(q, v)(t)$ and replace $q + L_q v$ by the value at $t = 1$ of the geodesic with initial value q and initial velocity $L_q v$. $\square$

In Lecture 30 we will need the following generalization of Theorem 8.4. For a proof see, for instance, either of [47, 58, 107].

Theorem 8.6. (*Coisotropic Embedding Theorem*) *Let M be a manifold of dimension $2n$, X a submanifold of dimension $k \geq n$, $i : X \hookrightarrow M$ the inclusion map, and ω_0 and ω_1 symplectic forms on M, such that $i^*\omega_0 = i^*\omega_1$ and X is coisotropic for both*

(M, ω_0) *and* (M, ω_1). *Then there exist neighborhoods* $\mathcal{U}_0$ *and* $\mathcal{U}_1$ *of X in M and a diffeomorphism* $\varphi : \mathcal{U}_0 \to \mathcal{U}_1$ *such that*

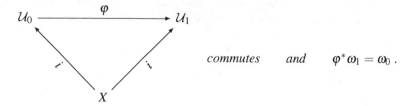

commutes   *and* $\varphi^* \omega_1 = \omega_0$.

Homework 6: Oriented Surfaces

1. The standard symplectic form on the 2-sphere is the standard area form:
   If we think of S^2 as the unit sphere in 3-space

   $$S^2 = \{u \in \mathbb{R}^3 \text{ such that } |u| = 1\} ,$$

   then the induced area form is given by

   $$\omega_u(v, w) = \langle u, v \times w \rangle$$

   where $u \in S^2$, $v, w \in T_u S^2$ are vectors in $\mathbb{R}^3$, $\times$ is the exterior product, and $\langle \cdot, \cdot \rangle$ is the standard inner product. With this form, the total area of S^2 is 4π.
   Consider cylindrical polar coordinates (θ, z) on S^2 away from its poles, where $0 \leq \theta < 2\pi$ and $-1 \leq z \leq 1$.
   Show that, in these coordinates,

   $$\omega = d\theta \wedge dz .$$

2. Prove the Darboux theorem in the 2-dimensional case, using the fact that every nonvanishing 1-form on a surface can be written locally as $f\, dg$ for suitable functions f, g.

   Hint: $\omega = df \wedge dg$ is nondegenerate $\iff$ (f, g) is a local diffeomorphism.

3. Any oriented 2-dimensional manifold with an area form is a symplectic manifold.

   (a) Show that convex combinations of two area forms ω_0, ω_1 that induce the same orientation are symplectic.

   This is wrong in dimension 4: find two symplectic forms on the vector space $\mathbb{R}^4$ that induce the same orientation, yet some convex combination of which is degenerate. Find a path of symplectic forms that connect them.

   (b) Suppose that we have two area forms ω_0, ω_1 on a compact 2-dimensional manifold M representing the same de Rham cohomology class, i.e., $[\omega_0] = [\omega_1] \in H^2_{\text{deRham}}(M)$.

   Prove that there is a 1-parameter family of diffeomorphisms $\varphi_t : M \to M$ such that $\varphi_1^* \omega_0 = \omega_1$, $\varphi_0 = \text{id}$, and $\varphi_t^* \omega_0$ is symplectic for all $t \in [0, 1]$.

   Hint: Exercise (a) and the Moser trick.

   Such a 1-parameter family φ_t is called a *strong isotopy* between ω_0 and ω_1. In this language, this exercise shows that, up to strong isotopy, there is a unique symplectic representative in each non-zero 2-cohomology class of M.

Chapter 9
Weinstein Tubular Neighborhood Theorem

9.1 Observation from Linear Algebra

Let (V, Ω) be a symplectic linear space, and let U be a lagrangian subspace.

Claim. There is a canonical nondegenerate bilinear pairing $\Omega' : V/U \times U \to \mathbb{R}$.

Proof. Define $\Omega'([v], u) = \Omega(v, u)$ where $[v]$ is the equivalence class of v in V/U. $\square$

Exercise. Check that Ω' is well-defined and nondegenerate. $\diamondsuit$

Consequently, we get
$\implies \widetilde{\Omega}' : V/U \to U^*$ defined by $\widetilde{\Omega}'([v]) = \Omega'([v], \cdot)$ is an isomorphism.
$\implies V/U \simeq U^*$ are canonically identified.

In particular, if (M, ω) is a symplectic manifold, and X is a lagrangian submanifold, then $T_x X$ is a lagrangian subspace of $(T_x M, \omega_x)$ for each $x \in X$.
The space $N_x X := T_x M / T_x X$ is called the **normal space** of X at x.

$\implies$ There is a canonical identification $N_x X \simeq T_x^* X$.
$\implies$

Theorem 9.1. *The vector bundles NX and T^*X are canonically identified.*

9.2 Tubular Neighborhoods

Theorem 9.2. (Standard Tubular Neighborhood Theorem) *Let M be an n-dimensional manifold, X a k-dimensional submanifold, NX the normal bundle of X in M, $i_0 : X \hookrightarrow NX$ the zero section, and $i : X \hookrightarrow M$ inclusion. Then there are neighborhoods $\mathcal{U}_0$ of X in NX, $\mathcal{U}$ of X in M and a diffeomorphism $\psi : \mathcal{U}_0 \to \mathcal{U}$ such that*

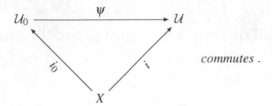

For the proof, see Lecture 6.

Theorem 9.3. (Weinstein Tubular Neighborhood Theorem) *Let (M,ω) be a symplectic manifold, X a compact lagrangian submanifold, ω_0 the canonical symplectic form on T^*X, $i_0 : X \hookrightarrow T^*X$ the lagrangian embedding as the zero section, and $i : X \hookrightarrow M$ the lagrangian embedding given by inclusion.*

Then there are neighborhoods $\mathcal{U}_0$ of X in T^*X, $\mathcal{U}$ of X in M, and a diffeomorphism $\varphi : \mathcal{U}_0 \to \mathcal{U}$ such that

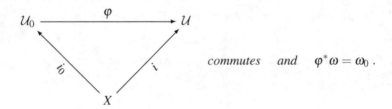

Proof. This proof relies on (1) the standard tubular neighborhood theorem, and (2) the Weinstein lagrangian neighborhood theorem.

(1) Since $NX \simeq T^*X$, we can find a neighborhood $\mathcal{N}_0$ of X in T^*X, a neighborhood $\mathcal{N}$ of X in M, and a diffeomorphism $\psi : \mathcal{N}_0 \to \mathcal{N}$ such that

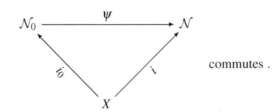

Let $\left. \begin{array}{l} \omega_0 = \text{canonical form on } T^*X \\ \omega_1 = \psi^*\omega \end{array} \right\}$ symplectic forms on $\mathcal{N}_0$.

The submanifold X is lagrangian for both ω_0 and ω_1.

(2) There exist neighborhoods $\mathcal{U}_0$ and $\mathcal{U}_1$ of X in $\mathcal{N}_0$ and a diffeomorphism $\theta : \mathcal{U}_0 \to \mathcal{U}_1$ such that

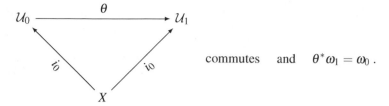

commutes　　and　$\theta^*\omega_1 = \omega_0$.

Take $\varphi = \psi \circ \theta$ and $\mathcal{U} = \varphi(\mathcal{U}_0)$. Check that $\varphi^*\omega = \theta^*\underbrace{\psi^*\omega}_{\omega_1} = \omega_0$.

□

Remark. Theorem 9.3 classifies lagrangian embeddings: up to local symplectomorphism, the set of lagrangian embeddings is the set of embeddings of manifolds into their cotangent bundles as zero sections.

The classification of *isotropic* embeddings was also carried out by Weinstein in [105, 107]. An **isotropic embedding** of a manifold X into a symplectic manifold (M, ω) is a closed embedding $i : X \hookrightarrow M$ such that $i^*\omega = 0$. Weinstein showed that neighbourhood equivalence of isotropic embeddings is in one-to-one correspondence with isomorphism classes of symplectic vector bundles.

The classification of *coisotropic embeddings* is due to Gotay [47]. A **coisotropic embedding** of a manifold X carrying a closed 2-form α of constant rank into a symplectic manifold (M, ω) is an embedding $i : X \hookrightarrow M$ such that $i^*\omega = \alpha$ and $i(X)$ is coisotropic as a submanifold of M. Let E be the **characteristic distribution** of a closed form α of constant rank on X, i.e., E_p is the kernel of α_p at $p \in X$. Gotay showed that then E^* carries a symplectic structure in a neighbourhood of the zero section, such that X embeds coisotropically onto this zero section, and, moreover every coisotropic embedding is equivalent to this in some neighbourhood of the zero section.　　　◇

9.3 Application 1: Tangent Space to the Group of Symplectomorphisms

The symplectomorphisms of a symplectic manifold (M, ω) form the group

$$\text{Sympl}(M, \omega) = \{f : M \overset{\simeq}{\longrightarrow} M \mid f^*\omega = \omega\} \, .$$

– What is $T_{\text{id}}(\text{Sympl}(M, \omega))$?
(What is the "Lie algebra" of the group of symplectomorphisms?)
– What does a neighborhood of id in $\text{Sympl}(M, \omega)$ look like?
We use notions from the C^1-topology:

C^1-**topology**.

Let X and Y be manifolds.

Definition 9.4. A sequence of maps $f_i : X \to Y$ *converges in the C^0-topology* to $f : X \to Y$ if and only if f_i converges uniformly on compact sets.

Definition 9.5. A sequence of C^1 maps $f_i : X \to Y$ *converges in the C^1-topology* to $f : X \to Y$ if and only if it and the sequence of derivatives $df_i : TX \to TY$ converge uniformly on compact sets.

Let (M, ω) be a compact symplectic manifold and $f \in \mathrm{Sympl}(M, \omega)$. Then

$$\left. \begin{array}{l} \text{Graph } f \\ \text{Graph id} = \Delta \end{array} \right\} \text{ are lagrangian submanifolds of } (M \times M, \mathrm{pr}_1^* \omega - \mathrm{pr}_2^* \omega).$$

($\mathrm{pr}_i : M \times M \to M$, $i = 1, 2$, are the projections to each factor.)

By the Weinstein tubular neighborhood theorem, there exists a neighborhood $\mathcal{U}$ of $\Delta\,(\simeq M)$ in $(M \times M, \mathrm{pr}_1^* \omega - \mathrm{pr}_2^* \omega)$ which is symplectomorphic to a neighborhood $\mathcal{U}_0$ of M in (T^*M, ω_0). Let $\varphi : \mathcal{U} \to \mathcal{U}_0$ be the symplectomorphism satisfying $\varphi(p, p) = (p, 0)$, $\forall p \in M$.

Suppose that f is sufficiently C^1-**close** to id, i.e., f is in some sufficiently small neighborhood of id in the C^1-topology. Then:

1. We can assume that Graph $f \subseteq \mathcal{U}$.

   Let $j : M \hookrightarrow \mathcal{U}$ be the embedding as Graph f,
   $i : M \hookrightarrow \mathcal{U}$ be the embedding as Graph id $= \Delta$.

2. The map j is sufficiently C^1-close to i.
3. By the Weinstein theorem, $\mathcal{U} \simeq \mathcal{U}_0 \subseteq T^*M$, so the above j and i induce

   $j_0 : M \hookrightarrow \mathcal{U}_0$ embedding, where $j_0 = \varphi \circ j$,
   $i_0 : M \hookrightarrow \mathcal{U}_0$ embedding as 0-section.

   Hence, we have

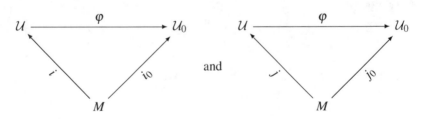

   where $i(p) = (p, p)$, $i_0(p) = (p, 0)$, $j(p) = (p, f(p))$ and $j_0(p) = \varphi(p, f(p))$ for $p \in M$.

4. The map j_0 is sufficiently C^1-close to i_0.

   $\Downarrow$

   The image set $j_0(M)$ intersects each $T_p^* M$ at one point μ_p depending smoothly on p.

5. The image of j_0 is the image of a smooth section $\mu : M \to T^*M$, that is, a 1-form $\mu = j_0 \circ (\pi \circ j_0)^{-1}$.

   Therefore, Graph $f \simeq \{(p, \mu_p) \mid p \in M,\ \mu_p \in T_p^* M\}$.

Exercise. Vice-versa: if μ is a 1-form sufficiently C^1-close to the zero 1-form, then

$$\{(p,\mu_p) \mid p \in M, \ \mu_p \in T_p^*M\} \ \simeq \ \text{Graph } f \ ,$$

for some diffeomorphism $f : M \rightarrow M$. By Lecture 3, we have

Graph f is lagrangian $\Longleftrightarrow$ μ is closed. $\qquad\qquad\qquad\qquad \diamondsuit$

Conclusion. A small C^1-neighborhood of id in $\text{Sympl}(M, \omega)$ is homeomorphic to a C^1-neighborhood of zero in the vector space of closed 1-forms on M. So:

$$T_{\text{id}}(\text{Sympl}(M, \omega)) \simeq \{\mu \in \Omega^1(M) \mid d\mu = 0\} \ .$$

In particular, $T_{\text{id}}(\text{Sympl}(M, \omega))$ contains the space of exact 1-forms

$$\{\mu = dh \mid h \in C^\infty(M)\} \ \simeq \ C^\infty(M)/ \text{ locally constant functions.}$$

9.4 Application 2: Fixed Points of Symplectomorphisms

Theorem 9.6. *Let* (M, ω) *be a compact symplectic manifold with* $H^1_{\text{deRham}}(M) = 0$. *Then any symplectomorphism of* M *which is sufficiently* C^1-*close to the identity has at least two fixed points.*

Proof. Suppose that $f \in \text{Sympl}(M, \omega)$ is sufficiently C^1-close to id.
   Then Graph $f \simeq$ closed 1-form μ on M.

$$\left. \begin{array}{l} d\mu = 0 \\ H^1_{\text{deRham}}(M) = 0 \end{array} \right\} \Longrightarrow \mu = dh \text{ for some } h \in C^\infty(M) \ .$$

Since M is compact, h has at least 2 critical points.

$$\begin{array}{ccc} \text{Fixed points of f} = & & \text{critical points of } h \\ \| & & \| \\ \text{Graph } f \cap \Delta & = & \{p : \mu_p = dh_p = 0\} \ . \end{array}$$

$\qquad\qquad\qquad\qquad\qquad\qquad\qquad\qquad\qquad\qquad\qquad\qquad\qquad\qquad\qquad \square$

Lagrangian intersection problem:
   A submanifold Y of M is C^1-**close** to X when there is a diffeomorphism $X \rightarrow Y$ which is, as a map into M, C^1-close to the inclusion $X \hookrightarrow M$.

Theorem 9.7. *Let* (M, ω) *be a symplectic manifold. Suppose that* X *is a compact lagrangian submanifold of* M *with* $H^1_{\text{deRham}}(X) = 0$. *Then every lagrangian submanifold of* M *which is* C^1-*close to* X *intersects* X *in at least two points.*

Proof. Exercise.   □

Arnold conjecture:

Let (M, ω) be a compact symplectic manifold, and $f : M \to M$ a symplectomor-phism which is "exactly homotopic to the identity" (see below). Then

$$\#\{\text{fixed points of } f\} \geq \text{minimal \# of critical points}$$
$$\text{a smooth function on } M \text{ can have.}$$

Together with Morse theory,[1] we obtain[2]

$$\#\{\text{nondegenerate fixed points of } f\} \geq \text{minimal \# of critical points}$$
$$\text{a Morse function on } M \text{ can have}$$
$$\geq \sum_{i=0}^{2n} \dim H^i(M; \mathbb{R}) \ .$$

The Arnold conjecture was proved by Conley-Zehnder, Floer, Hofer-Salamon, Ono, Fukaya-Ono, Liu-Tian using Floer homology (which is an ∞-dimensional ana-logue of Morse theory). There are open conjectures for sharper bounds on the num-ber of fixed points.

Meaning of "f is exactly homotopic to the identity:"

Suppose that $h_t : M \to \mathbb{R}$ is a smooth family of functions which is 1-periodic, i.e., $h_t = h_{t+1}$. Let $\rho : M \times \mathbb{R} \to M$ be the isotopy generated by the time-dependent vector field v_t defined by $\omega(v_t, \cdot) = dh_t$. Then "$f$ being exactly homotopic to the identity" means $f = \rho_1$ for some such h_t.

In other words, f is **exactly homotopic to the identity** when f is the time-1 map of an isotopy generated by some smooth time-dependent 1-periodic hamiltonian function.

There is a one-to-one correspondence

$$\text{fixed points of } f \quad \overset{1-1}{\longleftrightarrow} \quad \text{period-1 orbits of } \rho : M \times \mathbb{R} \to M$$

because $f(p) = p$ if and only if $\{\rho(t, p) , t \in [0, 1]\}$ is a closed orbit.

Proof of the Arnold conjecture in the case when $h : M \to \mathbb{R}$ is independent of t:
p is a critical point of $h \iff dh_p = 0 \iff v_p = 0$
$\implies \rho(t, p) = p , \forall t \in \mathbb{R} \implies p$ is a fixed point of ρ_1.   □

Exercise. Compute these estimates for the number of fixed points on some compact symplectic manifolds (for instance, S^2, $S^2 \times S^2$ and $T^2 = S^1 \times S^1$).   ◇

---

[1] A **Morse function** on M is a function $h : M \to \mathbb{R}$ whose critical points (i.e., points p where $dh_p = 0$) are all nondegenerate (i.e., the hessian at those points is nonsingular: $\det \left(\frac{\partial^2 h}{\partial x_i \partial x_j} \right)_p \neq 0$).

[2] A fixed point p of $f : M \to M$ is **nondegenerate** if $df_p : T_p M \to T_p M$ is nonsingular.

Part IV
Contact Manifolds

Contact geometry is also known as "the odd-dimensional analogue of symplectic geometry." We will browse through the basics of contact manifolds and their relation to symplectic manifolds.

Chapter 10
Contact Forms

10.1 Contact Structures

Definition 10.1. A *contact element* on a manifold M is a point $p \in M$, called the *contact point*, together with a tangent hyperplane at p, $H_p \subset T_pM$, that is, a codimension-1 subspace of T_pM.

A hyperplane $H_p \subset T_pM$ determines a covector $\alpha_p \in T_p^*M \setminus \{0\}$, up to multiplication by a nonzero scalar:

$$(p, H_p) \text{ is a contact element} \longleftrightarrow H_p = \ker \alpha_p \text{ with } \alpha_p : T_pM \longrightarrow \mathbb{R} \text{ linear}, \neq 0$$

$$\ker \alpha_p = \ker \alpha_p' \iff \alpha_p = \lambda \alpha_p' \text{ for some } \lambda \in \mathbb{R} \setminus \{0\} .$$

Suppose that H is a smooth field of contact elements (i.e., of tangent hyperplanes) on M:

$$H : p \longmapsto H_p \subset T_pM .$$

Locally, $H = \ker \alpha$ for some 1-form α, called a **locally defining 1-form** for H. (α is not unique: $\ker \alpha = \ker(f\alpha)$, for any nowhere vanishing $f : M \to \mathbb{R}$.)

Definition 10.2. A *contact structure* on M is a smooth field of tangent hyperplanes $H \subset TM$, such that, for any locally defining 1-form α, we have $d\alpha|_H$ nondegenerate (i.e., symplectic). The pair (M, H) is then called a *contact manifold* and α is called a *local contact form*.

At each $p \in M$,

$$T_pM - \underbrace{\ker \alpha_p}_{H_p} \oplus \underbrace{\ker d\alpha_p}_{1-\text{dimensional}} .$$

The $\ker d\alpha_p$ summand in this splitting depends on the choice of α.

$$d\alpha_p|_{H_p} \text{ nondegenerate} \implies \begin{cases} \dim H_p = 2n & \text{is even} \\ (d\alpha_p)^n|_{H_p} \neq 0 \text{ is a volume form on } H_p \end{cases}$$

$$\alpha_p|_{\ker d\alpha_p} \text{ nondegenerate}$$

Therefore,

- any contact manifold (M,H) has $\dim M = 2n+1$ *odd*, and
- if α is a (global) contact form, then $\alpha \wedge (d\alpha)^n$ is a volume form on M.

Remark. Let (M,H) be a contact manifold. A *global* contact form exists if and only if the quotient line bundle TM/H is orientable. Since H is also orientable, this implies that M is orientable.   ◇

Proposition 10.3. *Let H be a field of tangent hyperplanes on M. Then*

H *is a contact structure* $\iff \alpha \wedge (d\alpha)^n \neq 0$ *for every locally defining 1-form α .*

Proof.
$\implies$ Done above.
$\impliedby$ Suppose that $H = \ker\alpha$ locally. We need to show:

$$d\alpha|_H \text{ nondegenerate} \iff \alpha \wedge (d\alpha)^n \neq 0 .$$

Take a local trivialization $\{e_1, f_1, \ldots, e_n, f_n, r\}$ of $TM = \ker\alpha \oplus$ rest , such that $\ker\alpha = \mathrm{span}\{e_1, f_1, \ldots, e_n, f_n\}$ and rest $= \mathrm{span}\{r\}$.

$$(\alpha \wedge (d\alpha)^n)(e_1, f_1, \ldots, e_n, f_n, r) = \underbrace{\alpha(r)}_{\neq 0} \cdot (d\alpha)^n(e_1, f_1, \ldots, e_n, f_n)$$

and hence $\alpha \wedge (d\alpha)^n \neq 0 \iff (d\alpha)^n|_H \neq 0 \iff d\alpha|_H$ is nondegenerate .   □

10.2 Examples

1. On $\mathbb{R}^3$ with coordinates (x, y, z), consider $\alpha = x\,dy + dz$. Since

$$\alpha \wedge d\alpha = (x\,dy + dz) \wedge (dx \wedge dy) = dx \wedge dy \wedge dz \neq 0 ,$$

α is a contact form on $\mathbb{R}^3$.
The corresponding field of hyperplanes $H = \ker\alpha$ at $(x, y, z) \in \mathbb{R}^3$ is

$$H_{(x,y,z)} = \{v = a\frac{\partial}{\partial x} + b\frac{\partial}{\partial y} + c\frac{\partial}{\partial z} \mid \alpha(v) = bx + c = 0\} .$$

Exercise. Picture these hyperplanes.   ◇

2. **(Martinet [80], 1971)** Any compact orientable 3-manifold admits a contact structure.
Open Problem, 2000. The classification of compact orientable contact 3-manifolds is still not known. There is by now a huge collection of results in

contact topology related to the classification of contact manifolds. For a review of the state of the knowledge and interesting questions on contact 3-manifolds, see [33, 43, 100].

3. Let X be a manifold and T^*X its cotangent bundle. There are two canonical contact manifolds associated to X (see Homework 7):

$$\mathbb{P}(T^*X) = \text{the projectivization of } T^*X, \text{ and}$$
$$S(T^*X) = \text{the cotangent sphere bundle.}$$

4. On $\mathbb{R}^{2n+1}$ with coordinates $(x_1, y_1, \ldots, x_n, y_n, z)$, $\alpha = \sum_i x_i dy_i + dz$ is contact.

10.3 First Properties

There is a local normal form theorem for contact manifolds analogous to the Darboux theorem for symplectic manifolds.

Theorem 10.4. *Let (M, H) be a contact manifold and $p \in M$. Then there exists a coordinate system $(\mathcal{U}, x_1, y_1, \ldots, x_n, y_n, z)$ centered at p such that on $\mathcal{U}$*

$$\alpha = \sum x_i dy_i + dz \text{ is a local contact form for } H.$$

The idea behind the proof is sketched in the next lecture.

There is also a Moser-type theorem for contact forms.

Theorem 10.5. (Gray) *Let M be a compact manifold. Suppose that α_t, $t \in [0, 1]$, is a smooth family of (global) contact forms on M. Let $H_t = \ker \alpha_t$. Then there exists an isotopy $\rho : M \times \mathbb{R} \longrightarrow M$ such that $H_t = \rho_{t*} H_0$, for all $0 \leq t \leq 1$.*

Exercise. Show that $H_t = \rho_{t*} H_0 \iff \rho_t^* \alpha_t = u_t \cdot \alpha_0$ for some family $u_t : M \longrightarrow \mathbb{R}$, $0 \leq t \leq 1$, of nowhere vanishing functions. $\qquad \diamond$

Proof. (À la Moser)

We need to find ρ_t such that $\begin{cases} \rho_0 = \text{id} \\ \frac{d}{dt}(\rho_t^* \alpha_t) = \frac{d}{dt}(u_t \alpha_0) . \end{cases}$ For any isotopy ρ,

$$\frac{d}{dt}(\rho_t^* \alpha_t) = \rho_t^* \left(\mathcal{L}_{v_t} \alpha_t + \frac{d\alpha_t}{dt} \right) ,$$

where $v_t = \frac{d\rho_t}{dt} \circ \rho_t^{-1}$ is the vector field generated by ρ_t. By the Moser trick, it suffices to find v_t and then integrate it to ρ_t. We will search for v_t in $H_t = \ker \alpha_t$; this unnecessary assumption simplifies the proof.

We need to solve

$$\rho_t^* (\underbrace{\mathcal{L}_{v_t} \alpha_t}_{d\iota_{v_t} \alpha_t + \iota_{v_t} d\alpha_t} + \frac{d\alpha_t}{dt}) = \frac{du_t}{dt} \underbrace{\alpha_0}_{\frac{1}{u_t} \rho_t^* \alpha_t}$$

$$\Longrightarrow \qquad \rho_t^* \left(\iota_{v_t} d\alpha_t + \frac{d\alpha_t}{dt} \right) = \frac{du_t}{dt} \cdot \frac{1}{u_t} \cdot \rho_t^* \alpha_t$$

$$\Longleftrightarrow \qquad \iota_{v_t} d\alpha_t + \frac{d\alpha_t}{dt} = (\rho_t^*)^{-1} \left(\frac{du_t}{dt} \cdot \frac{1}{u_t} \right) \alpha_t . \qquad (\star)$$

Restricting to the hyperplane $H_t = \ker \alpha_t$, equation $(\star)$ reads

$$\iota_{v_t} d\alpha_t |_{H_t} = - \frac{d\alpha_t}{dt} \Big|_{H_t}$$

which determines v_t uniquely, since $d\alpha_t |_{H_t}$ is nondegenerate. After integrating v_t to ρ_t, the factor u_t is determined by the relation $\rho_t^* \alpha_t = u_t \cdot \alpha_0$. Check that this indeed gives a solution. $\square$

Homework 7: Manifolds of Contact Elements

Given any manifold X of dimension n, there is a canonical symplectic manifold of dimension $2n$ attached to it, namely its cotangent bundle with the standard symplectic structure. The exercises below show that there is also a canonical *contact* manifold of dimension $2n - 1$ attached to X.

The **manifold of contact elements** of an n-dimensional manifold X is

$$\mathcal{C} = \{(x, \chi_x) \mid x \in X \text{ and } \chi_x \text{ is a hyperplane in } T_x X\} \ .$$

On the other hand, the projectivization of the cotangent bundle of X is

$$\mathbb{P}^* X = (T^* X \setminus \text{zero section}) / \sim$$

where $(x, \xi) \sim (x, \xi')$ whenever $\xi = \lambda \xi'$ for some $\lambda \in \mathbb{R} \setminus \{0\}$ (here $x \in X$ and $\xi, \xi' \in T_x^* X \setminus \{0\}$). We will denote elements of $\mathbb{P}^* X$ by $(x, [\xi])$, $[\xi]$ being the $\sim$ equivalence class of ξ.

1. Show that $\mathcal{C}$ is naturally isomorphic to $\mathbb{P}^* X$ as a bundle over X, i.e., exhibit a diffeomorphism $\varphi : \mathcal{C} \to \mathbb{P}^* X$ such that the following diagram commutes:

$$
\begin{array}{ccc}
\mathcal{C} & \xrightarrow{\varphi} & \mathbb{P}^* X \\
\pi \downarrow & & \downarrow \pi \\
X & = & X
\end{array}
$$

   where the vertical maps are the natural projections $(x, \chi_x) \mapsto x$ and $(x, \xi) \mapsto x$.

   Hint: The kernel of a non-zero $\xi \in T_x^* X$ is a hyperplane $\chi_x \subset T_x X$. What is the relation between ξ and ξ' if $\ker \xi = \ker \xi'$?

2. There is on $\mathcal{C}$ a canonical field of hyperplanes $\mathcal{H}$ (that is, a smooth map attaching to each point in $\mathcal{C}$ a hyperplane in the tangent space to $\mathcal{C}$ at that point): $\mathcal{H}$ at the point $p = (x, \chi_x) \in \mathcal{C}$ is the hyperplane

$$\mathcal{H}_p = (d\pi_p)^{-1} \chi_x \subset T_p \mathcal{C} \ ,$$

   where

$$
\begin{array}{ccc}
\mathcal{C} & p = (x, \chi_x) & T_p \mathcal{C} \\
\downarrow \pi & \downarrow & \downarrow d\pi_p \\
X & x & T_x X
\end{array}
$$

   are the natural projections, and $(d\pi_p)^{-1} \chi_x$ is the preimage of $\chi_x \subset T_x X$ by $d\pi_p$. Under the isomorphism $\mathcal{C} \simeq \mathbb{P}^* X$ from exercise 1, $\mathcal{H}$ induces a field of hyperplanes $\mathbb{H}$ on $\mathbb{P}^* X$. Describe $\mathbb{H}$.

   Hint: If $\xi \in T_x^* X \setminus \{0\}$ has kernel χ_x, what is the kernel of the canonical 1-form $\alpha_{(x, \xi)} = (d\pi_{(x, \xi)})^* \xi$?

3. Check that $(\mathbb{P}^*X, \mathbb{H})$ is a contact manifold, and therefore $(\mathcal{C}, \mathcal{H})$ is a contact manifold.

>   **Hint:** Let $(x, [\xi]) \in \mathbb{P}^*X$. For any ξ representing the class $[\xi]$, we have
>
> $$\mathbb{H}_{(x.[\xi])} = \ker\left((d\pi_{(x.[\xi])})^*\xi\right) .$$
>
>   Let $x_1, \ldots, x_n$ be local coordinates on X, and let $x_1, \ldots, x_n, \xi_1, \ldots, \xi_n$ be the associated local coordinates on T^*X. In these coordinates, $(x, [\xi])$ is given by $(x_1, \ldots, x_n, [\xi_1, \ldots, \xi_n])$. Since at least one of the ξ_i's is nonzero, without loss of generality we may assume that $\xi_1 \neq 0$ so that we may divide ξ by ξ_1 to obtain a representative with coordinates $(1, \xi_2, \ldots, \xi_n)$. Hence, by choosing always the representative of $[\xi]$ with $\xi_1 = 1$, the set $x_1, \ldots, x_n, \xi_2, \ldots, \xi_n$ defines coordinates on some neighborhood $\mathcal{U}$ of $(x, [\xi])$ in $\mathbb{P}^*X$. On $\mathcal{U}$, consider the 1-form
>
> $$\alpha = dx_1 + \sum_{i \geq 2} \xi_i dx_i .$$
>
>   Show that α is a contact form on $\mathcal{U}$, i.e., show that $\ker \alpha_{(x.[\xi])} = \mathbb{H}_{(x.[\xi])}$, and that $d\alpha_{(x.[\xi])}$ is nondegenerate on $\mathbb{H}_{(x.[\xi])}$.

4. What is the symplectization of $\mathcal{C}$?
   What is the manifold $\mathcal{C}$ when $X = \mathbb{R}^3$ and when $X = S^1 \times S^1$?

Remark. Similarly, we could have defined the **manifold of oriented contact elements** of X to be

$$\mathcal{C}^o = \left\{ (x, \chi_x^o) \,\middle|\, x \in X \text{ and } \begin{array}{l} \chi_x^o \text{ is a hyperplane in } T_xX \\ \text{equipped with an orientation} \end{array} \right\} .$$

The manifold $\mathcal{C}^o$ is isomorphic to the cotangent sphere bundle of X

$$S^*X := (T^*X \setminus \text{zero section}) / \approx$$

where $(x, \xi) \approx (x, \xi')$ whenever $\xi = \lambda \xi'$ for some $\lambda \in \mathbb{R}^+$.
A construction analogous to the above produces a canonical contact structure on $\mathcal{C}^o$. See [3, Appendix 4].

$$\Diamond$$

Chapter 11
Contact Dynamics

11.1 Reeb Vector Fields

Let (M, H) be a contact manifold with a contact form α.

Claim. There exists a unique vector field R on M such that $\begin{cases} \iota_R d\alpha = 0 \\ \iota_R \alpha = 1 \end{cases}$

Proof. $\begin{cases} \iota_R d\alpha = 0 \implies R \in \ker d\alpha \text{, which is a line bundle, and} \\ \iota_R \alpha = 1 \implies \text{normalizes } R \text{.} \end{cases}$ $\qquad \square$

The vector field R is called the **Reeb vector field** determined by α.

Claim. The flow of R preserves the contact form, i.e., if $\rho_t = \exp tR$ is the isotopy generated by R, then $\rho_t^* \alpha = \alpha, \forall t \in \mathbb{R}$.

Proof. We have $\frac{d}{dt}(\rho_t^* \alpha) = \rho_t^*(\mathcal{L}_R \alpha) = \rho_t^*(d \underbrace{\iota_R \alpha}_{1} + \underbrace{\iota_R d\alpha}_{0}) = 0$.

Hence, $\rho_t^* \alpha = \rho_0^* \alpha = \alpha, \forall t \in \mathbb{R}$. $\qquad \square$

Definition 11.1. A *contactomorphism* is a diffeomorphism f of a contact manifold (M, H) which preserves the contact structure (i.e., $f_* H = H$).

Examples.

1. Euclidean space $\mathbb{R}^{2n+1}$ with $\alpha = \sum_i x_i dy_i + dz$.

$$\left. \begin{array}{l} \iota_R \sum dx_i \wedge dy_i = 0 \\ \iota_R \sum x_i dy_i + dz = 1 \end{array} \right\} \implies R = \frac{\partial}{\partial z} \text{ is the Reeb vector field.}$$

The contactomorphisms generated by R are translations

$$\rho_t(x_1, y_1, \ldots, x_n, y_n, z) = (x_1, y_1, \ldots, x_n, y_n, z + t) .$$

2. Regard the odd sphere $S^{2n-1} \xrightarrow{i} \mathbb{R}^{2n}$ as the set of unit vectors

$$\{(x_1, y_1, \ldots, x_n, y_n) \mid \sum(x_i^2 + y_i^2) = 1\}.$$

Consider the 1-form on $\mathbb{R}^{2n}$, $\sigma = \frac{1}{2}\sum(x_i dy_i - y_i dx_i)$.

Claim. The form $\alpha = i^*\sigma$ is a contact form on S^{2n-1}.

Proof. We need to show that $\alpha \wedge (d\alpha)^{n-1} \neq 0$. The 1-form on $\mathbb{R}^{2n}$ $v = d\sum(x_i^2 + y_i^2) = 2\sum(x_i dx_i + y_i dy_i)$ satisfies $T_p S^{2n-1} = \ker v_p$, at $p \in S^{2n-1}$. Check that $v \wedge \sigma \wedge (d\sigma)^{n-1} \neq 0$. $\qquad\square$

The distribution $H = \ker\alpha$ is called the **standard contact structure** on S^{2n-1}. The Reeb vector field is $R = 2\sum\left(x_i\frac{\partial}{\partial y_i} - y_i\frac{\partial}{\partial x_i}\right)$, and is also known as the **Hopf vector field** on S^{2n-1}, as the orbits of its flow are the circles of the Hopf fibration.

$\diamondsuit$

11.2 Symplectization

Example. Let $\widetilde{M} = S^{2n-1} \times \mathbb{R}$, with coordinate τ in the $\mathbb{R}$-factor, and projection $\pi : \widetilde{M} \to S^{2n-1}$, $(p, \tau) \mapsto p$. Under the identification $\widetilde{M} \simeq \mathbb{R}^{2n}\setminus\{0\}$, where the $\mathbb{R}$-factor represents the logarithm of the square of the radius, the projection π becomes

$$\pi : \quad \mathbb{R}^{2n}\setminus\{0\} \quad \longrightarrow \quad S^{2n-1}$$
$$(X_1, Y_1, \ldots, X_n, Y_n) \longmapsto (\tfrac{X_1}{\sqrt{e^\tau}}, \tfrac{Y_1}{\sqrt{e^\tau}}, \ldots, \tfrac{X_n}{\sqrt{e^\tau}}, \tfrac{Y_n}{\sqrt{e^\tau}})$$

where $e^\tau = \sum(X_i^2 + Y_i^2)$. Let $\alpha = i^*\sigma$ be the standard contact form on S^{2n-1} (see the previous example). Then $\omega = d(e^\tau\pi^*\alpha)$ is a closed 2-form on $\mathbb{R}^{2n}\setminus\{0\}$. Since $\pi^*i^*x_i = \frac{X_i}{\sqrt{e^\tau}}$, $\pi^*i^*y_i = \frac{Y_i}{\sqrt{e^\tau}}$, we have

$$\pi^*\alpha = \pi^*i^*\sigma = \frac{1}{2}\sum\left(\tfrac{X_i}{\sqrt{e^\tau}}d(\tfrac{Y_i}{\sqrt{e^\tau}}) - \tfrac{Y_i}{\sqrt{e^\tau}}d(\tfrac{X_i}{\sqrt{e^\tau}})\right)$$
$$= \frac{1}{2e^\tau}\sum(X_i dY_i - Y_i dX_i).$$

Therefore, $\omega = \sum dX_i \wedge dY_i$ is the standard symplectic form on $\mathbb{R}^{2n}\setminus\{0\} \subset \mathbb{R}^{2n}$. $(\widetilde{M}, \omega)$ is called the *symplectization* of (S^{2n-1}, α). $\diamondsuit$

Proposition 11.2. *Let (M, H) be a contact manifold with a contact form α. Let $\widetilde{M} = M \times \mathbb{R}$, and let $\pi : \widetilde{M} \to M$, $(p, \tau) \mapsto p$, be the projection. Then $\omega = d(e^\tau\pi^*\alpha)$ is a symplectic form on $\widetilde{M}$, where τ is a coordinate on $\mathbb{R}$.*

Proof. Exercise. □

Hence, $\widetilde{M}$ has a symplectic form ω canonically determined by a contact form α on M and a coordinate function on $\mathbb{R}$; $(\widetilde{M}, \omega)$ is called the **symplectization** of (M, α).

Remarks.

1. The contact version of the Darboux theorem can now be derived by applying the symplectic theorem to the symplectization of the contact manifold (with appropriate choice of coordinates); see [3, Appendix 4].
2. There is a coordinate-free description of $\widetilde{M}$ as

$$\widetilde{M} = \{(p, \xi) \mid p \in M,\ \xi \in T_p^* M,\ \text{such that}\ \ker \xi = H_p\} .$$

The group $\mathbb{R} \setminus \{0\}$ acts on $\widetilde{M}$ by multiplication on the cotangent vector:

$$\lambda \cdot (p, \xi) = (p, \lambda \xi), \quad \lambda \in \mathbb{R} \setminus \{0\} .$$

The quotient $\widetilde{M}/(\mathbb{R} \setminus \{0\})$ is diffeomorphic to M. $\widetilde{M}$ has a canonical 1-form $\widetilde{\alpha}$ defined at $v \in T_{(p,\xi)}\widetilde{M}$ by

$$\widetilde{\alpha}_{(p,\xi)}(v) = \xi\big((d\,\mathrm{pr})_{(p,\xi)}v\big) ,$$

where $\mathrm{pr} : \widetilde{M} \to M$ is the bundle projection.

$\diamondsuit$

11.3 Conjectures of Seifert and Weinstein

Question. (Seifert, 1948) Let v be a nowhere vanishing vector field on the 3-sphere. Does the flow of v have any periodic orbits?

Counterexamples.

- **(Schweitzer, 1974)** $\exists C^1$ vector field without periodic orbits.
- **(Kristina Kuperberg, 1994)** $\exists C^\infty$ vector field without periodic orbits.

Question. How about volume-preserving vector fields?

- **(Greg Kuperberg, 1997)** $\exists C^1$ counterexample.
- C^∞ counterexamples are not known.

Natural generalization of this problem:

Let $M = S^3$ be the 3-sphere, and let γ be a volume form on M. Suppose that v is a nowhere vanishing vector field, and suppose that v is volume-preserving, i.e.,

$$\mathcal{L}_v \gamma = 0 \iff d\iota_v \gamma = 0 \iff \iota_v \gamma = d\alpha$$

for some 1-form α, since $H^2(S^3) = 0$.

Given a 1-form α, we would like to study vector fields v such that

$$\begin{cases} \iota_v \gamma = d\alpha \\ \iota_v \alpha > 0 \,. \end{cases}$$

A vector field v satisfying $\iota_v \alpha > 0$ is called **positive**. For instance, vector fields in a neighborhood of the Hopf vector field are positive relative to the standard contact form on S^3.

Renormalizing as $R := \frac{v}{\iota_v \alpha}$, we should study instead

$$\begin{cases} \iota_R d\alpha = 0 \\ \iota_\alpha = 1 \\ \alpha \wedge d\alpha \text{ is a volume form,} \end{cases}$$

that is, study pairs (α, R) where

$$\begin{cases} \alpha \text{ is a \textbf{contact} form, and} \\ R \text{ is its \textbf{Reeb} vector field.} \end{cases}$$

Conjecture. *(Weinstein, 1978 [106]) Suppose that M is a 3-dimensional manifold with a (global) contact form α. Let v be the Reeb vector field for α. Then v has a periodic orbit.*

Theorem 11.3. *(Viterbo and Hofer, 1993 [63, 64, 103])   The Weinstein conjecture is true when*

1) $M = S^3$, or
2) $\pi_2(M) \neq 0$, or
3) the contact structure is overtwisted.[1]

---

[1] A surface S inside a contact 3-manifold determines a singular foliation on S, called the **characteristic foliation** of S, by the intersection of the contact planes with the tangent spaces to S. A contact structure on a 3-manifold M is called **overtwisted** if there exists an embedded 2-disk whose characteristic foliation contains one closed leaf C and exactly one singular point inside C; otherwise, the contact structure is called **tight**. Eliashberg [32] showed that the isotopy classification of overtwisted contact structures on closed 3-manifolds coincides with their homotopy classification as tangent plane fields. The classification of tight contact structures is still open.

Open questions.

- How many periodic orbits are there?
- What do they look like?
- Is there always an unknotted one?
- What about the linking behavior?

Part V
Compatible Almost Complex Structures

The fact that any symplectic manifold possesses almost complex structures, and even so in a *compatible* sense, establishes a link from symplectic geometry to complex geometry, and is the point of departure for the modern technique of counting pseudo-holomorphic curves, as first proposed by Gromov [49].

Chapter 12
Almost Complex Structures

12.1 Three Geometries

1. Symplectic geometry:
   geometry of a closed nondegenerate skew-symmetric bilinear form.
2. Riemannian geometry:
   geometry of a positive-definite symmetric bilinear map.
3. Complex geometry:
   geometry of a linear map with square -1.

Example. The euclidean space $\mathbb{R}^{2n}$ with the standard linear coordinates $(x_1,\ldots,x_n,y_1,\ldots,y_n)$ has standard structures:

$$\omega_0 = \sum dx_j \wedge dy_j \text{ , standard symplectic structure;}$$

$$g_0 = \langle \cdot,\cdot \rangle , \qquad \text{standard inner product; and}$$

if we identify $\mathbb{R}^{2n}$ with $\mathbb{C}^n$ with coordinates $z_j = x_j + \sqrt{-1}\,y_j$, then multiplication by $\sqrt{-1}$ induces a constant linear map J_0 on the tangent spaces of $\mathbb{R}^{2n}$:

$$J_0(\frac{\partial}{\partial x_j}) = \frac{\partial}{\partial y_j} , \qquad J_0(\frac{\partial}{\partial y_j}) = -\frac{\partial}{\partial x_j} ,$$

with $J_0^2 = -\mathrm{Id}$. Relative to the basis $\frac{\partial}{\partial x_1},\ldots,\frac{\partial}{\partial x_n},\frac{\partial}{\partial y_1},\ldots,\frac{\partial}{\partial y_n}$, the maps J_0, ω_0 and g_0 are represented by

$$J_0(u) = \begin{pmatrix} 0 & -\mathrm{Id} \\ \mathrm{Id} & 0 \end{pmatrix} u$$

$$\omega_0(u,v) = v^t \begin{pmatrix} 0 & -\mathrm{Id} \\ \mathrm{Id} & 0 \end{pmatrix} u$$

$$g_0(u,v) = v^t u$$

where $u, v \in \mathbb{R}^{2n}$ and v^t is the transpose of v. The following compatibility relation holds:

$$\omega_0(u, v) = g_0(J_0(u), v) \ .$$

12.2 Complex Structures on Vector Spaces

Definition 12.1. Let V be a vector space. A ***complex structure*** on V is a linear map:

$$J : V \to V \qquad \text{with} \qquad J^2 = -\text{Id} \ .$$

The pair (V, J) is called a ***complex vector space***.

A complex structure J is equivalent to a structure of vector space over $\mathbb{C}$ if we identify the map J with multiplication by $\sqrt{-1}$.

Definition 12.2. Let (V, Ω) be a symplectic vector space. A complex structure J on V is said to be ***compatible*** (with Ω, or Ω-compatible) if

$$G_J(u, v) := \Omega(u, Jv) \ , \quad \forall u, v \in V \ , \text{ is a positive inner product on } V \ .$$

That is,

$$J \text{ is } \Omega\text{-compatible} \iff \begin{cases} \Omega(Ju, Jv) = \Omega(u, v) & [\text{symplectomorphism}] \\ \Omega(u, Ju) > 0, \quad \forall u \neq 0 & [\text{taming condition}] \end{cases}$$

Compatible complex structures always exist on symplectic vector spaces:

Proposition 12.3. *Let (V, Ω) be a symplectic vector space. Then there is a compatible complex structure J on V.*

Proof. Choose a positive inner product G on V. Since Ω and G are nondegenerate,

$$\left. \begin{array}{l} u \in V \longmapsto \Omega(u, \cdot) \in V^* \\ w \in V \longmapsto G(w, \cdot) \in V^* \end{array} \right\} \text{ are isomorphisms between } V \text{ and } V^*.$$

Hence, $\Omega(u, v) = G(Au, v)$ for some linear map $A : V \to V$. This map A is skew-symmetric because

$$\begin{aligned} G(A^* u, v) = G(u, Av) &= G(Av, u) \\ &= \Omega(v, u) = -\Omega(u, v) = G(-Au, v) \ . \end{aligned}$$

Also:

- AA^* is symmetric: $(AA^*)^* = AA^*$.
- AA^* is positive: $G(AA^* u, u) = G(A^* u, A^* u) > 0$, for $u \neq 0$.

These properties imply that AA^* diagonalizes with positive eigenvalues λ_i,

$$AA^* = B \operatorname{diag}(\lambda_1, \ldots, \lambda_{2n}) B^{-1} .$$

We may hence define an arbitrary real power of AA^* by rescaling the eigenspaces, in particular,

$$\sqrt{AA^*} := B \operatorname{diag}(\sqrt{\lambda_1}, \ldots, \sqrt{\lambda_{2n}}) B^{-1} .$$

Then $\sqrt{AA^*}$ is symmetric and positive-definite. Let

$$J = (\sqrt{AA^*})^{-1} A .$$

The factorization $A = \sqrt{AA^*} J$ is called the **polar decomposition** of A. Since A commutes with $\sqrt{AA^*}$, J commutes with $\sqrt{AA^*}$. Check that J is orthogonal, $JJ^* = \operatorname{Id}$, as well as skew-adjoint, $J^* = -J$, and hence it is a complex structure on V:

$$J^2 = -JJ^* = -\operatorname{Id} .$$

Compatibility:

$$\Omega(Ju, Jv) = G(AJu, Jv) = G(JAu, Jv) = G(Au, v)$$
$$= \Omega(u, v)$$
$$\Omega(u, Ju) = G(Au, Ju) = G(-JAu, u)$$
$$= G(\sqrt{AA^*} u, u) > 0 , \quad \text{for } u \neq 0 .$$

Therefore, J is a compatible complex structure on V. $\qquad\square$

As indicated in the proof, in general, the positive inner product defined by

$$\Omega(u, Jv) = G(\sqrt{AA^*} u, v) \text{ is different from } G(u, v) .$$

Remarks.

1. This construction is canonical after an initial choice of G. To see this, notice that $\sqrt{AA^*}$ does not depend on the choice of B nor of the ordering of the eigenvalues in $\operatorname{diag}(\sqrt{\lambda_1}, \ldots, \sqrt{\lambda_{2n}})$. The linear transformation $\sqrt{AA^*}$ is completely determined by its effect on each eigenspace of AA^*: on the eigenspace corresponding to the eigenvalue λ_k, the map $\sqrt{AA^*}$ is defined to be multiplication by $\sqrt{\lambda_k}$.

2. If (V_t, Ω_t) is a family of symplectic vector spaces with a family G_t of positive inner products, all depending smoothly on a real parameter t, then, adapting the proof of the previous proposition, we can show that there is a smooth family J_t of compatible complex structures on V_t.

3. To check just the existence of compatible complex structures on a symplectic vector space (V, Ω), we could also proceed as follows. Given a symplectic basis $e_1, \ldots, e_n, f_1, \ldots, f_n$ (i.e., $\Omega(e_i, e_j) = \Omega(f_i, f_j) = 0$ and $\Omega(e_i, f_j) = \delta_{ij}$), one can define $Je_j = f_j$ and $Jf_j = -e_j$. This is a compatible complex structure on (V, Ω).

Moreover, given Ω and J compatible on V, there exists a symplectic basis of V of the form:

$$e_1, \ldots, e_n, f_1 = Je_1, \ldots, f_n = Je_n \ .$$

The proof is part of Homework 8.

4. Conversely, given (V, J), there is always a symplectic structure Ω such that J is Ω-compatible: pick any positive inner product G such that $J^* = -J$ and take $\Omega(u, v) = G(Ju, v)$.

12.3 Compatible Structures

Definition 12.4. An *almost complex structure* on a manifold M is a smooth field of complex structures on the tangent spaces:

$$x \longmapsto J_x : T_x M \to T_x M \quad \text{linear}, \quad \text{and} \quad J_x^2 = -\text{Id} \ .$$

The pair (M, J) is then called an *almost complex manifold*.

Definition 12.5. Let (M, ω) be a symplectic manifold. An almost complex structure J on M is called *compatible* (with ω or ω-compatible) if the assignment

$$x \longmapsto g_x : T_x M \times T_x M \to \mathbb{R}$$
$$g_x(u, v) := \omega_x(u, J_x v)$$

is a riemannian metric on M.

For a manifold M,

ω is a symplectic form $\implies x \longmapsto \omega_x : T_x M \times T_x M \to \mathbb{R}$ is bilinear, nondegenerate, skew-symmetric;

g is a riemannian metric $\implies x \longmapsto g_x : T_x M \times T_x M \to \mathbb{R}$ is a positive inner product;

J almost complex structure $\implies x \longmapsto J_x : T_x M \to T_x M$ is linear and $J^2 = -\text{Id}$.

The triple (ω, g, J) is called a **compatible triple** when $g(\cdot, \cdot) = \omega(\cdot, J \cdot)$.

Proposition 12.6. *Let (M, ω) be a symplectic manifold, and g a riemannian metric on M. Then there exists a canonical almost complex structure J on M which is compatible.*

Proof. The polar decomposition is *canonical* (after a choice of metric), hence this construction of J on M is *smooth*; cf. Remark 2 of the previous section. $\qquad \square$

Remark. In general, $g_J(\cdot,\cdot) := \omega(\cdot,J\cdot) \neq g(\cdot,\cdot)$. $\Diamond$

Since riemannian metrics always exist, we conclude:

Corollary 12.7. *Any symplectic manifold has compatible almost complex structures.*

– How different can compatible almost complex structures be?

Proposition 12.8. *Let (M,ω) be a symplectic manifold, and J_0, J_1 two almost complex structures compatible with ω. Then there is a smooth family $J_t, 0 \leq t \leq 1$, of compatible almost complex structures joining J_0 to J_1.*

Proof. By compatibility, we get

$$\left.\begin{array}{l} \omega, J_0 \rightsquigarrow g_0(\cdot,\cdot) = \omega(\cdot,J_0\cdot) \\ \omega, J_1 \rightsquigarrow g_1(\cdot,\cdot) = \omega(\cdot,J_1\cdot) \end{array}\right\} \quad \text{two riemannian metrics on } M \,.$$

Their convex combinations

$$g_t(\cdot,\cdot) = (1-t)g_0(\cdot,\cdot) + tg_1(\cdot,\cdot) \,, \qquad 0 \leq t \leq 1 \,,$$

form a smooth family of riemannian metrics. Apply the polar decomposition to (ω, g_t) to obtain a smooth family of J_t's joining J_0 to J_1. $\square$

Corollary 12.9. *The set of all compatible almost complex structures on a symplectic manifold is path-connected.*

Homework 8: Compatible Linear Structures

1. Let $\Omega(V)$ and $J(V)$ be the spaces of symplectic forms and complex structures on the vector space V, respectively. Take $\Omega \in \Omega(V)$ and $J \in J(V)$. Let $GL(V)$ be the group of all isomorphisms of V, let $Sp(V,\Omega)$ be the group of symplecto-morphisms of (V,Ω), and let $GL(V,J)$ be the group of complex isomorphisms of (V,J).
   Show that
   $$\Omega(V) \simeq GL(V)/Sp(V,\Omega) \quad \text{and} \quad J(V) \simeq GL(V)/GL(V,J) .$$

   Hint: The group $GL(V)$ acts on $\Omega(V)$ by pullback. What is the stabilizer of a given Ω?

2. Let $(\mathbb{R}^{2n},\Omega_0)$ be the standard $2n$-dimensional symplectic euclidean space. The **symplectic linear group** is the group of all linear transformations of $\mathbb{R}^{2n}$ which preserve the symplectic structure:
   $$Sp(2n) := \{A \in GL(2n;\mathbb{R}) \mid \Omega_0(Au,Av) = \Omega_0(u,v) \text{ for all } u,v \in \mathbb{R}^{2n}\} .$$

   Identifying the complex $n \times n$ matrix $X + iY$ with the real $2n \times 2n$ matrix $\begin{pmatrix} X & -Y \\ Y & X \end{pmatrix}$, consider the following subgroups of $GL(2n;\mathbb{R})$:
   $$Sp(2n) , O(2n) , GL(n;\mathbb{C}) \text{ and } U(n) .$$

   Show that the intersection of any two of them is $U(n)$. (From [83, p.41].)

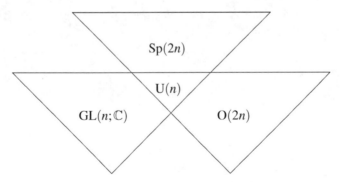

3. Let (V,Ω) be a symplectic vector space of dimension $2n$, and let $J : V \to V$, $J^2 = -\text{Id}$, be a complex structure on V.

   (a) Prove that, if J is Ω-compatible and L is a lagrangian subspace of (V,Ω), then JL is also lagrangian and $JL = L^{\perp}$, where $\perp$ denotes orthogonality with respect to the positive inner product $G_J(u,v) = \Omega(u,Jv)$.
   (b) Deduce that J is Ω-compatible if and only if there exists a symplectic basis for V of the form
   $$e_1,e_2,\ldots,e_n, f_1 = Je_1, f_2 = Je_2,\ldots, f_n = Je_n$$
   where $\Omega(e_i,e_j) = \Omega(f_i,f_j) = 0$ and $\Omega(e_i,f_j) = \delta_{ij}$.

Chapter 13
Compatible Triples

13.1 Compatibility

Let (M, ω) be a symplectic manifold. As shown in the previous lecture, compatible almost complex structures always exist on (M, ω). We also showed that the set of all compatible almost complex structures on (M, ω) is path-connected. In fact, the set of all compatible almost complex structures is even contractible. (This is important for defining invariants.) Let $\mathcal{J}(T_x M, \omega_x)$ be the set of all compatible complex structures on $(T_x M, \omega_x)$ for $x \in M$.

Proposition 13.1. *The set $\mathcal{J}(T_x M, \omega_x)$ is contractible, i.e., there exists a homotopy*

$$h_t : \mathcal{J}(T_x M, \omega_x) \longrightarrow \mathcal{J}(T_x M, \omega_x) , \quad 0 \leq t \leq 1 ,$$

starting at the identity $h_0 = \mathrm{Id}$,
finishing at a trivial map $h_1 : \mathcal{J}(T_x M, \omega_x) \to \{J_0\}$,
and fixing J_0 (i.e., $h_t(J_0) = J_0$, $\forall t$) for some $J_0 \in \mathcal{J}(T_x M, \omega_x)$.

Proof. Homework 9. □

Consider the fiber bundle $\mathcal{J} \to M$ with fiber

$$\mathcal{J}_x := \mathcal{J}(T_x M, \omega_x) \quad \text{over } x \in M .$$

A compatible almost complex structure J on (M, ω) is a section of $\mathcal{J}$. The space of sections of $\mathcal{J}$ is contractible because the fibers are contractible.

Remarks.

- We never used the closedness of ω to construct compatible almost complex structures. The construction holds for an **almost symplectic manifold** (M, ω), that is, a pair of a manifold M and a nondegenerate 2-form ω, not necessarily closed.

- Similarly, we could define a **symplectic vector bundle** to be a vector bundle $E \rightarrow M$ over a manifold M equipped with a smooth field ω of fiberwise nondegenerate skew-symmetric bilinear maps

$$\omega_x : E_x \times E_x \longrightarrow \mathbb{R} \ .$$

The existence of such a field ω is equivalent to being able to reduce the structure group of the bundle from the general linear group to the linear symplectic group. As a consequence of our discussion, a symplectic vector bundle is always a complex vector bundle, and vice-versa.

13.2 Triple of Structures

If (ω, J, g) is a **compatible triple**, then any one of ω, J or g can be written in terms of the other two:

$$g(u,v) = \omega(u, Jv)$$
$$\omega(u,v) = g(Ju, v)$$
$$J(u) = \widetilde{g}^{-1}(\widetilde{\omega}(u))$$

where

$$\widetilde{\omega} : TM \longrightarrow T^*M \qquad u \longmapsto \omega(u, \cdot)$$
$$\widetilde{g} : TM \longrightarrow T^*M \qquad u \longmapsto g(u, \cdot)$$

are the linear isomorphisms induced by the bilinear forms ω and g.

The relations among ω, J and g can be summarized in the following table. The last column lists differential equations these structures are usually asked to satisfy.

Data	Condition/Technique	Consequence	Question
ω, J	$\omega(Ju, Jv) = \omega(u, v)$ $\omega(u, Ju) > 0, u \neq 0$	$g(u,v) := \omega(u, Jv)$ is positive inner product	(g flat?)
g, J	$g(Ju, Jv) = g(u, v)$ (i.e., J is orthogonal)	$\omega(u,v) := g(Ju, v)$ is nondeg., skew-symm.	ω closed?
ω, g	polar decomposition $\rightsquigarrow$	J almost complex str.	J integrable?

An almost complex structure J on a manifold M is called **integrable** if and only if J is induced by a structure of complex manifold on M. In Lecture 15 we will discuss tests to check whether a given J is integrable.

13.3 First Consequences

Proposition 13.2. *Let (M,J) be an almost complex manifold. Suppose that J is compatible with two symplectic structures ω_0, ω_1 Then ω_0, ω_1 are deformation-equivalent, that is, there exists a smooth family ω_t, $0 \leq t \leq 1$, of symplectic forms joining ω_0 to ω_1.*

Proof. Take $\omega_t = (1-t)\omega_0 + t\omega_1$, $0 \leq t \leq 1$. Then:

- ω_t is closed.
- ω_t is nondegenerate, since

$$g_t(\cdot, \cdot) := \omega_t(\cdot, J\cdot) = (1-t)g_0(\cdot, \cdot) + tg_1(\cdot, \cdot)$$

is positive, hence nondegenerate.

$\square$

Remark. The converse of this proposition is not true. A counterexample is provided by the following family in $\mathbb{R}^4$:

$$\omega_t = \cos \pi t \, dx_1 dy_1 + \sin \pi t \, dx_1 dy_2 + \sin \pi t \, dy_1 dx_2 + \cos \pi t \, dx_2 dy_2 \,, 0 \leq t \leq 1 \,.$$

There is no J in $\mathbb{R}^4$ compatible with both ω_0 and ω_1. $\diamondsuit$

Definition 13.3. A submanifold X of an almost complex manifold (M,J) is an **almost complex submanifold** when $J(TX) \subseteq TX$, i.e., for all $x \in X, v \in T_xX$, we have $J_xv \in T_xX$.

Proposition 13.4. *Let (M, ω) be a symplectic manifold equipped with a compatible almost complex structure J. Then any almost complex submanifold X of (M,J) is a symplectic submanifold of (M, ω).*

Proof. Let $i : X \hookrightarrow M$ be the inclusion. Then $i^*\omega$ is a closed 2-form on X. Nondegeneracy:

$$\omega_x(u, v) = g_x(J_xu, v) \,, \qquad \forall x \in X \,, \forall u, v \in T_xX \,.$$

Since $g_x|_{T_xX}$ is nondegenerate, so is $\omega_x|_{T_xX}$. Hence, $i^*\omega$ is symplectic. $\square$

– When is an almost complex manifold a complex manifold? See Lecture 15.

Examples.

S^2 is an almost complex manifold and it is a complex manifold.
S^4 is not an almost complex manifold (proved by Ehresmann and Hopf).
S^6 is almost complex and it is not yet known whether it is complex.
S^8 and higher spheres are not almost complex manifolds.

$\diamondsuit$

Homework 9: Contractibility

The following proof illustrates in a geometric way the relation between lagrangian subspaces, complex structures and inner products; from [11, p.45].

Let (V, Ω) be a symplectic vector space, and let $\mathcal{J}(V, \Omega)$ be the set of all complex structures on (V, Ω) which are Ω-compatible; i.e., given a complex structure J on V we have

$$J \in \mathcal{J}(V, \Omega) \iff G_J(\cdot, \cdot) := \Omega(\cdot, J\cdot) \text{ is a positive inner product on } V .$$

Fix a lagrangian subspace L_0 of (V, Ω). Let $\mathcal{L}(V, \Omega, L_0)$ be the space of all lagrangian subspaces of (V, Ω) which intersect L_0 transversally. Let $\mathcal{G}(L_0)$ be the space of all positive inner products on L_0.

Consider the map

$$\Psi : \mathcal{J}(V, \Omega) \to \mathcal{L}(V, \Omega, L_0) \times \mathcal{G}(L_0)$$
$$J \mapsto (JL_0, G_J|_{L_0})$$

Show that:

1. Ψ is well-defined.
2. Ψ is a bijection.

> **Hint:** Given $(L, G) \in \mathcal{L}(V, \Omega, L_0) \times \mathcal{G}(L_0)$, define J in the following manner:
> For $v \in L_0$, $v^{\perp} = \{u \in L_0 \mid G(u, v) = 0\}$ is a $(n-1)$-dimensional space of L_0; its symplectic orthogonal $(v^{\perp})^{\Omega}$ is $(n+1)$-dimensional. Check that $(v^{\perp})^{\Omega} \cap L$ is 1-dimensional. Let Jv be the unique vector in this line such that $\Omega(v, Jv) = 1$. Check that, if we take v's in some G-orthonormal basis of L_0, this defines the required element of $\mathcal{J}(V, \Omega)$.

3. $\mathcal{L}(V, \Omega, L_0)$ is contractible.

> **Hint:** Prove that $\mathcal{L}(V, \Omega, L_0)$ can be identified with the vector space of all symmetric $n \times n$ matrices. Notice that any n-dimensional subspace L of V which is transversal to L_0 is the graph of a linear map $S : JL_0 \to L_0$, i.e.,
>
> $$L = \text{span of } \{Je_1 + SJe_1, \ldots, Je_n + SJe_n\}$$
> $$\text{when} \quad L_0 = \text{span of } \{e_1, \ldots, e_n\} .$$

4. $\mathcal{G}(L_0)$ is contractible.

> **Hint:** $\mathcal{G}(L_0)$ is even convex.

Conclude that $\mathcal{J}(V, \Omega)$ is contractible.

Chapter 14
Dolbeault Theory

14.1 Splittings

Let (M, J) be an almost complex manifold. The complexified tangent bundle of M is the bundle

$$TM \otimes \mathbb{C}$$
$$\downarrow$$
$$M$$

with fiber $(TM \otimes \mathbb{C})_p = T_p M \otimes \mathbb{C}$ at $p \in M$. If

$T_p M$ is a $2n$-dimensional vector space over $\mathbb{R}$, then
$T_p M \otimes \mathbb{C}$ is a $2n$-dimensional vector space over $\mathbb{C}$.

We may extend J linearly to $TM \otimes \mathbb{C}$:

$$J(v \otimes c) = Jv \otimes c , \quad v \in TM , \quad c \in \mathbb{C} .$$

Since $J^2 = -\mathrm{Id}$, on the complex vector space $(TM \otimes \mathbb{C})_p$, the linear map J_p has eigenvalues $\pm i$. Let

$$
\begin{aligned}
T_{1,0} &= \{ v \in TM \otimes \mathbb{C} \mid Jv = +iv \} = (+i)\text{-eigenspace of } J \\
&= \{ v \otimes 1 - Jv \otimes i \mid v \in TM \} \\
&= (J\text{-})\textbf{holomorphic tangent vectors} ;
\end{aligned}
$$

$$
\begin{aligned}
T_{0,1} &= \{ v \in TM \otimes \mathbb{C} \mid Jv = -iv \} = (-i)\text{-eigenspace of } J \\
&= \{ v \otimes 1 + Jv \otimes i \mid v \in TM \} \\
&= (J\text{-})\textbf{anti-holomorphic tangent vectors} .
\end{aligned}
$$

Since

$$
\begin{aligned}
\pi_{1,0} : TM &\longrightarrow T_{1,0} \\
v &\longmapsto \tfrac{1}{2}(v \otimes 1 - Jv \otimes i)
\end{aligned}
$$

is a (real) bundle isomorphism such that $\pi_{1,0} \circ J = i\pi_{1,0}$, and

$$\pi_{0,1} : TM \longrightarrow T_{0,1}$$
$$v \longmapsto \tfrac{1}{2}(v \otimes 1 + Jv \otimes i)$$

is also a (real) bundle isomorphism such that $\pi_{0,1} \circ J = -i\pi_{0,1}$, we conclude that we have isomorphisms of complex vector bundles

$$(TM, J) \simeq T_{1,0} \simeq \overline{T_{0,1}} \ ,$$

where $\overline{T_{0,1}}$ denotes the complex conjugate bundle of $T_{0,1}$. Extending $\pi_{1,0}$ and $\pi_{0,1}$ to projections of $TM \otimes \mathbb{C}$, we obtain an isomorphism

$$(\pi_{1,0}, \pi_{0,1}) : TM \otimes \mathbb{C} \xrightarrow{\simeq} T_{1,0} \oplus T_{0,1} \ .$$

Similarly, the complexified cotangent bundle splits as

$$(\pi^{1,0}, \pi^{0,1}) : T^*M \otimes \mathbb{C} \xrightarrow{\simeq} T^{1,0} \oplus T^{0,1}$$

where

$$T^{1,0} = (T_{1,0})^* = \{\eta \in T^* \otimes \mathbb{C} \mid \eta(J\omega) = i\eta(\omega)\, , \forall \omega \in TM \otimes \mathbb{C}\}$$
$$= \{\xi \otimes 1 - (\xi \circ J) \otimes i \mid \xi \in T^*M\}$$
$$= \textbf{complex-linear cotangent vectors}\ ,$$

$$T^{0,1} = (T_{0,1})^* = \{\eta \in T^* \otimes \mathbb{C} \mid \eta(J\omega) = -i\eta(\omega)\, , \forall \omega \in TM \otimes \mathbb{C}\}$$
$$= \{\xi \otimes 1 + (\xi \circ J) \otimes i \mid \xi \in T^*M\}$$
$$= \textbf{complex-antilinear cotangent vectors}\ ,$$

and $\pi^{1,0}, \pi^{0,1}$ are the two natural projections

$$\pi^{1,0} : T^*M \otimes \mathbb{C} \longrightarrow T^{1,0}$$
$$\eta \longmapsto \eta^{1,0} := \tfrac{1}{2}(\eta - i\eta \circ J) \ ;$$

$$\pi^{0,1} : T^*M \otimes \mathbb{C} \longrightarrow T^{0,1}$$
$$\eta \longmapsto \eta^{0,1} := \tfrac{1}{2}(\eta + i\eta \circ J) \ .$$

14.2 Forms of Type (ℓ, m)

For an almost complex manifold (M, J), let

$$\Omega^k(M; \mathbb{C}) := \text{sections of } \Lambda^k(T^*M \otimes \mathbb{C})$$
$$= \textbf{complex-valued k-forms on } M, where$$

$$\Lambda^k(T^*M \otimes \mathbb{C}) := \Lambda^k(T^{1,0} \oplus T^{0,1})$$
$$= \oplus_{\ell+m=k} \underbrace{(\Lambda^\ell T^{1,0}) \wedge (\Lambda^m T^{0,1})}_{\Lambda^{\ell,m}(\text{definition})}$$
$$= \oplus_{\ell+m=k} \Lambda^{\ell,m} .$$

In particular, $\Lambda^{1,0} = T^{1,0}$ and $\Lambda^{0,1} = T^{0,1}$.

Definition 14.1. The *differential forms of type* (ℓ,m) on (M,J) are the sections of $\Lambda^{\ell,m}$:

$$\Omega^{\ell,m} := \text{ sections of } \Lambda^{\ell,m} .$$

Then

$$\Omega^k(M;\mathbb{C}) = \oplus_{\ell+m=k}\Omega^{\ell,m} .$$

Let $\pi^{\ell,m} : \Lambda^k(T^*M \otimes \mathbb{C}) \to \Lambda^{\ell,m}$ be the projection map, where $\ell+m = k$. The usual exterior derivative d composed with two of these projections induces differential operators ∂ and $\bar{\partial}$ on forms of type (ℓ,m):

$$\partial := \pi^{\ell+1,m} \circ d : \Omega^{\ell,m}(M) \longrightarrow \Omega^{\ell+1,m}(M)$$
$$\bar{\partial} := \pi^{\ell,m+1} \circ d : \Omega^{\ell,m}(M) \longrightarrow \Omega^{\ell,m+1}(M) .$$

If $\beta \in \Omega^{\ell,m}(M)$, with $k = \ell+m$, then $d\beta \in \Omega^{k+1}(M;\mathbb{C})$:

$$d\beta = \sum_{r+s=k+1} \pi^{r,s}d\beta = \pi^{k+1,0}d\beta + \cdots + \partial\beta + \bar{\partial}\beta + \cdots + \pi^{0,k+1}d\beta .$$

14.3 *J*-Holomorphic Functions

Let $f : M \to \mathbb{C}$ be a smooth complex-valued function on M. The exterior derivative d extends linearly to $\mathbb{C}$-valued functions as $df = d(\text{Re}f) + id(\text{Im}f)$.

Definition 14.2. A function f is *(J-)holomorphic at* $x \in M$ if df_p is complex linear, i.e., $df_p \circ J = i df_p$. A function f is *(J-)holomorphic* if it is holomorphic at all $p \in M$.

Exercise. Show that

$$df_p \circ J = i df_p \quad \Longleftrightarrow \quad df_p \in T_p^{1,0} \quad \Longleftrightarrow \quad \pi_p^{0,1}df_p = 0 .$$

$\Diamond$

Definition 14.3. A function f is *(J-)anti-holomorphic at* $p \in M$ if df_p is complex antilinear, i.e., $df_p \circ J = -i df_p$.

Exercise.

$$df_p \circ J = -i df_p \iff df_p \in T_p^{0,1} \iff \pi_p^{1,0} df_p = 0$$
$$\iff d\bar{f}_p \in T_p^{1,0} \iff \pi_p^{0,1} d\bar{f}_p = 0$$
$$\iff \bar{f} \text{ is holomorphic at } p \in M .$$

$\diamondsuit$

Definition 14.4. On functions, $d = \partial + \bar\partial$, where

$$\partial := \pi^{1,0} \circ d \quad \text{and} \quad \bar\partial := \pi^{0,1} \circ d .$$

Then

$$f \text{ is holomorphic} \iff \bar\partial f = 0 ,$$
$$f \text{ is anti-holomorphic} \iff \partial f = 0 .$$

– What about higher differential forms?

14.4 Dolbeault Cohomology

Suppose that $d = \partial + \bar\partial$, i.e.,

$$d\beta = \underbrace{\partial\beta}_{\in \Omega^{\ell+1,m}} + \underbrace{\bar\partial\beta}_{\in \Omega^{\ell,m+1}} , \quad \forall \beta \in \Omega^{\ell,m} .$$

Then, for any form $\beta \in \Omega^{\ell,m}$,

$$0 = d^2\beta = \underbrace{\partial^2\beta}_{\in \Omega^{\ell+2,m}} + \underbrace{\partial\bar\partial\beta + \bar\partial\partial\beta}_{\in \Omega^{\ell+1,m+1}} + \underbrace{\bar\partial^2\beta}_{\in \Omega^{\ell,m+2}} ,$$

which implies

$$\begin{cases} \bar\partial^2 = 0 \\ \partial\bar\partial + \bar\partial\partial = 0 \\ \partial^2 = 0 \end{cases}$$

Since $\bar\partial^2 = 0$, the chain

$$0 \longrightarrow \Omega^{\ell,0} \xrightarrow{\bar\partial} \Omega^{\ell,1} \xrightarrow{\bar\partial} \Omega^{\ell,2} \xrightarrow{\bar\partial} \cdots$$

is a differential complex; its cohomology groups

$$H_{\text{Dolbeault}}^{\ell,m}(M) := \frac{\ker \bar\partial : \Omega^{\ell,m} \longrightarrow \Omega^{\ell,m+1}}{\operatorname{im} \bar\partial : \Omega^{\ell,m-1} \longrightarrow \Omega^{\ell,m}}$$

are called the **Dolbeault cohomology** groups.

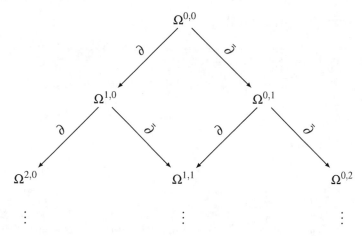

– When is $d = \partial + \bar{\partial}$? See the next lecture.

Homework 10: Integrability

This set of problems is from [11, p.46-47].

1. Let (M, J) be an almost complex manifold. Its **Nijenhuis tensor** $\mathcal{N}$ is:

$$\mathcal{N}(v, w) := [Jv, Jw] - J[v, Jw] - J[Jv, w] - [v, w] \, ,$$

   where v and w are vector fields on M, $[\cdot, \cdot]$ is the usual bracket

$$[v, w] \cdot f := v \cdot (w \cdot f) - w \cdot (v \cdot f) \, , \quad \text{for } f \in C^{\infty}(M) \, ,$$

   and $v \cdot f = df(v)$.

   (a) Check that, if the map $v \mapsto [v, w]$ is complex linear (in the sense that it commutes with J), then $\mathcal{N} \equiv 0$.
   (b) Show that $\mathcal{N}$ is actually a tensor, that is: $\mathcal{N}(v, w)$ at $x \in M$ depends only on the values $v_x, w_x \in T_x M$ and not really on the vector fields v and w.
   (c) Compute $\mathcal{N}(v, Jv)$. Deduce that, if M is a surface, then $\mathcal{N} \equiv 0$.

   A theorem of Newlander and Nirenberg [89] states that an almost complex manifold (M, J) is a complex (analytic) manifold if and only if $\mathcal{N} \equiv 0$. Combining (c) with the fact that any orientable surface is symplectic, we conclude that any orientable surface is a complex manifold, a result already known to Gauss.

2. Let $\mathcal{N}$ be as above. For any map $f : \mathbb{R}^{2n} \to \mathbb{C}$ and any vector field v on $\mathbb{R}^{2n}$, we have $v \cdot f = v \cdot (f_1 + i f_2) = v \cdot f_1 + i v \cdot f_2$, so that $f \mapsto v \cdot f$ is a complex linear map.

   (a) Let $\mathbb{R}^{2n}$ be endowed with an almost complex structure J, and suppose that f is a **J-holomorphic function**, that is,

$$df \circ J = i \, df \, .$$

   Show that $df(\mathcal{N}(v, w)) = 0$ for all vector fields v, w.
   (b) Suppose that there exist n J-holomorphic functions, $f_1, \ldots, f_n$, on $\mathbb{R}^{2n}$, which are independent at some point p, i.e., the real and imaginary parts of $(df_1)_p, \ldots, (df_n)_p$ form a basis of $T_p^* \mathbb{R}^{2n}$. Show that $\mathcal{N}$ vanishes identically at p.
   (c) Assume that M is a complex manifold and J is its complex structure. Show that $\mathcal{N}$ vanishes identically everywhere on M.

   In general, an almost complex manifold has *no* J-holomorphic functions at all. On the other hand, it has *plenty* of **J-holomorphic curves**: maps $f : \mathbb{C} \to M$ such that $df \circ i = J \circ df$. J-holomorphic curves, also known as **pseudo-holomorphic curves**, provide a main tool in symplectic topology, as first realized by Gromov [49].

Part VI
Kähler Manifolds

Kähler geometry lies at the intersection of complex, riemannian and symplectic geometries, and plays a central role in all of these fields. We will start by reviewing complex manifolds. After describing the local normal form for Kähler manifolds (Lecture 16), we conclude with a summary of Hodge theory for compact Kähler manifolds (Lecture 17).

Chapter 15
Complex Manifolds

15.1 Complex Charts

Definition 15.1. A *complex manifold* of (complex) dimension n is a set M with a complete complex atlas

$$\mathcal{A} = \{(\mathcal{U}_\alpha, \mathcal{V}_\alpha, \varphi_\alpha), \alpha \in \text{index set } I\}$$

where $M = \cup_\alpha \mathcal{U}_\alpha$, the $\mathcal{V}_\alpha$'s are open subsets of $\mathbb{C}^n$, and the maps $\varphi_\alpha : \mathcal{U}_\alpha \to \mathcal{V}_\alpha$ are such that the transition maps $\psi_{\alpha\beta}$ are *biholomorphic* as maps on open subsets of $\mathbb{C}^n$:

where $\mathcal{V}_{\alpha\beta} = \varphi_\alpha(\mathcal{U}_\alpha \cap \mathcal{U}_\beta) \subseteq \mathbb{C}^n$ and $\mathcal{V}_{\beta\alpha} = \varphi_\beta(\mathcal{U}_\alpha \cap \mathcal{U}_\beta) \subseteq \mathbb{C}^n$. $\psi_{\alpha\beta}$ being biholomorphic means that $\psi_{\alpha\beta}$ is a bijection and that $\psi_{\alpha\beta}$ and $\psi_{\alpha\beta}^{-1}$ are both holomorphic.

Proposition 15.2. *Any complex manifold has a canonical almost complex structure.*

Proof.

1) Local definition of J:

Let $(\mathcal{U}, \mathcal{V}, \varphi \cdot \mathcal{U} \to \mathcal{V})$ be a complex chart for a complex manifold M with $\varphi = (z_1, \ldots, z_n)$ written in components relative to complex coordinates $z_j = x_j + iy_j$. At $p \in \mathcal{U}$

$$T_p M = \mathbb{R}\text{-span of } \left\{ \frac{\partial}{\partial x_j}\bigg|_p, \frac{\partial}{\partial y_j}\bigg|_p : j = 1, \ldots, n \right\}.$$

Define J over $\mathcal{U}$ by

$$J_p\left(\left.\frac{\partial}{\partial x_j}\right|_p\right) = \left.\frac{\partial}{\partial y_j}\right|_p$$

$$j=1,\ldots,n\,.$$

$$J_p\left(\left.\frac{\partial}{\partial y_j}\right|_p\right) = -\left.\frac{\partial}{\partial x_j}\right|_p$$

2) *This J is well-defined globally*:

If $(\mathcal{U},\mathcal{V},\varphi)$ and $(\mathcal{U}',\mathcal{V}',\varphi')$ are two charts, we need to show that $J=J'$ on their overlap.

On $\mathcal{U}\cap\mathcal{U}'$, $\psi\circ\varphi=\varphi'$. If $z_j=x_j+iy_j$ and $w_j=u_j+iv_j$ are coordinates on $\mathcal{U}$ and $\mathcal{U}'$, respectively, so that φ and φ' can be written in components $\varphi=(z_1,\ldots,z_n)$, $\varphi'=(w_1,\ldots,w_n)$, then $\psi(z_1,\ldots,z_n)=(w_1,\ldots,w_n)$. Taking the derivative of a composition

$$\begin{cases} \dfrac{\partial}{\partial x_k} = \sum_j\left(\dfrac{\partial u_j}{\partial x_k}\dfrac{\partial}{\partial u_j}+\dfrac{\partial v_j}{\partial x_k}\dfrac{\partial}{\partial v_j}\right)\\[2ex] \dfrac{\partial}{\partial y_k} = \sum_j\left(\dfrac{\partial u_j}{\partial y_k}\dfrac{\partial}{\partial u_j}+\dfrac{\partial v_j}{\partial y_k}\dfrac{\partial}{\partial v_j}\right) \end{cases}$$

Since ψ is biholomorphic, each component of ψ satisfies the **Cauchy-Riemann equations**:

$$\begin{cases} \dfrac{\partial u_j}{\partial x_k} = \dfrac{\partial v_j}{\partial y_k}\\[2ex] \dfrac{\partial u_j}{\partial y_k} = -\dfrac{\partial v_j}{\partial x_k} \end{cases} \qquad j,k=1,\ldots,n\,.$$

These equations imply

$$J'\underbrace{\sum_j\left(\frac{\partial u_j}{\partial x_k}\frac{\partial}{\partial u_j}+\frac{\partial v_j}{\partial x_k}\frac{\partial}{\partial v_j}\right)}_{} = \sum_j\left(\frac{\partial u_j}{\partial y_k}\frac{\partial}{\partial u_j}+\frac{\partial v_j}{\partial y_k}\frac{\partial}{\partial v_j}\right)$$

$$\sum_j\left(\underbrace{\frac{\partial u_j}{\partial x_k}}_{\frac{\partial v_j}{\partial y_k}}\frac{\partial}{\partial v_j}-\underbrace{\frac{\partial v_j}{\partial x_j}}_{-\frac{\partial u_j}{\partial y_k}}\frac{\partial}{\partial u_j}\right)$$

which matches the equation

$$J\frac{\partial}{\partial x_k} = \frac{\partial}{\partial y_k}\,.$$

$\square$

15.2 Forms on Complex Manifolds

Suppose that M is a complex manifold and J is its canonical almost complex structure. What does the splitting $\Omega^k(M;\mathbb{C}) = \oplus_{\ell+m=k}\Omega^{\ell,m}$ look like? ([22, 48, 66, 109] are good references for this material.)

Let $\mathcal{U} \subseteq M$ be a coordinate neighborhood with complex coordinates $z_1,\ldots,z_n$, $z_j = x_j + iy_j$, and real coordinates $x_1,y_1,\ldots,x_n,y_n$. At $p \in \mathcal{U}$,

$$T_pM = \mathbb{R}\text{-span}\left\{\left.\frac{\partial}{\partial x_j}\right|_p, \left.\frac{\partial}{\partial y_j}\right|_p\right\}$$

$$T_pM \otimes \mathbb{C} = \mathbb{C}\text{-span}\left\{\left.\frac{\partial}{\partial x_j}\right|_p, \left.\frac{\partial}{\partial y_j}\right|_p\right\}$$

$$= \underbrace{\mathbb{C}\text{-span}\left\{\frac{1}{2}\left(\left.\frac{\partial}{\partial x_j}\right|_p - i\left.\frac{\partial}{\partial y_j}\right|_p\right)\right\}}_{\substack{T_{1,0} = (+i)\text{-eigenspace of } J \\ J\left(\frac{\partial}{\partial x_j} - i\frac{\partial}{\partial y_j}\right) = i\left(\frac{\partial}{\partial x_j} - i\frac{\partial}{\partial y_j}\right)}} \oplus \underbrace{\mathbb{C}\text{-span}\left\{\frac{1}{2}\left(\left.\frac{\partial}{\partial x_j}\right|_p + i\left.\frac{\partial}{\partial y_j}\right|_p\right)\right\}}_{\substack{T_{0,1} = (-i)\text{-eigenspace of } J \\ J\left(\frac{\partial}{\partial x_j} + i\frac{\partial}{\partial y_j}\right) = -i\left(\frac{\partial}{\partial x_j} + i\frac{\partial}{\partial y_j}\right)}}$$

This can be written more concisely using:

Definition 15.3.

$$\frac{\partial}{\partial z_j} := \frac{1}{2}\left(\frac{\partial}{\partial x_j} - i\frac{\partial}{\partial y_j}\right) \quad \text{and} \quad \frac{\partial}{\partial \bar{z}_j} := \frac{1}{2}\left(\frac{\partial}{\partial x_j} + i\frac{\partial}{\partial y_j}\right).$$

Hence,

$$(T_{1,0})_p = \mathbb{C}\text{-span}\left\{\left.\frac{\partial}{\partial z_j}\right|_p : j = 1,\ldots,n\right\}, \ (T_{0,1})_p = \mathbb{C}\text{-span}\left\{\left.\frac{\partial}{\partial \bar{z}_j}\right|_p : j = 1,\ldots,n\right\}.$$

Similarly,

$$T^*M \otimes \mathbb{C} = \mathbb{C}\text{-span}\{dx_j, dy_j : j = 1,\ldots,n\}$$

$$= \underbrace{\mathbb{C}\text{-span}\{dx_j + idy_j : j = 1,\ldots,n\}}_{T^{1,0}} \oplus \underbrace{\mathbb{C}\text{-span}\{dx_j - idy_j : j = 1,\ldots,n\}}_{T^{0,1}}$$

$$(dx_j + idy_j) \circ J = i(dx_j + idy_j) \qquad (dx_j - idy_j) \circ J = -i(dx_j - idy_j)$$

Putting

$$dz_j = dx_j + idy_j \qquad \text{and} \qquad d\bar{z}_j = dx_j - idy_j,$$

we obtain

$$T^{1,0} = \mathbb{C}\text{-span}\{dz_j : j = 1,\ldots,n\}, \quad T^{0,1} = \mathbb{C}\text{-span}\{d\bar{z}_j : j = 1,\ldots,n\}.$$

On the coordinate neighborhood $\mathcal{U}$,

$$
\begin{aligned}
(1,0)\text{-forms} &= \left\{ \sum_j b_j dz_j \mid b_j \in C^\infty(\mathcal{U};\mathbb{C}) \right\} \\
(0,1)\text{-forms} &= \left\{ \sum_j b_j d\bar{z}_j \mid b_j \in C^\infty(\mathcal{U};\mathbb{C}) \right\} \\
(2,0)\text{-forms} &= \left\{ \sum_{j_1<j_2} b_{j_1,j_2} dz_{j_1} \wedge dz_{j_2} \mid b_{j_1,j_2} \in C^\infty(\mathcal{U};\mathbb{C}) \right\} \\
(1,1)\text{-forms} &= \left\{ \sum_{j_1,j_2} b_{j_1,j_2} dz_{j_1} \wedge d\bar{z}_{j_2} \mid b_{j_1,j_2} \in C^\infty(\mathcal{U};\mathbb{C}) \right\} \\
(0,2)\text{-forms} &= \left\{ \sum_{j_1<j_2} b_{j_1,j_2} d\bar{z}_{j_1} \wedge d\bar{z}_{j_2} \mid b_{j_1,j_2} \in C^\infty(\mathcal{U};\mathbb{C}) \right\}
\end{aligned}
$$

If we use multi-index notation:

$$
\begin{aligned}
J &= (j_1,\ldots,j_m) \qquad\qquad 1 \le j_1 < \ldots < j_m \le n \\
|J| &= m \\
dz_J &= dz_{j_1} \wedge dz_{j_2} \wedge \ldots \wedge dz_{j_m}
\end{aligned}
$$

then

$$
\Omega^{\ell,m} = (\ell,m)\text{-forms} = \left\{ \sum_{|J|=\ell,|K|=m} b_{J,K} dz_J \wedge d\bar{z}_K \mid b_{J,K} \in C^\infty(\mathcal{U};\mathbb{C}) \right\}.
$$

15.3 Differentials

On a coordinate neighborhood $\mathcal{U}$, a form $\beta \in \Omega^k(M;\mathbb{C})$ may be written as

$$
\beta = \sum_{|J|+|K|=k} a_{J,K} dx_J \wedge dy_K, \qquad \text{with } a_{J,K} \in C^\infty(\mathcal{U};\mathbb{C}).
$$

We would like to know whether the following equality holds:

$$
d\beta = \sum (\partial a_{J,K} + \bar{\partial} a_{J,K}) dx_J \wedge dy_K \overset{?}{=} (\partial + \bar{\partial}) \sum a_{J,K} dx_J \wedge dy_K.
$$

If we use the identities

$$
\begin{cases} dx_j + i\,dy_j = dz_j \\ dx_j - i\,dy_j = d\bar{z}_j \end{cases} \quad\Longleftrightarrow\quad \begin{cases} dx_j = \frac{1}{2}(dz_j + d\bar{z}_j) \\ dy_j = \frac{1}{2i}(dz_j - d\bar{z}_j) \end{cases}
$$

after substituting and reshuffling, we obtain

$$
\beta = \sum_{|J|+|K|=k} b_{J,K} dz_J \wedge d\bar{z}_K
$$

$$
= \sum_{\ell+m=k} \underbrace{\left(\sum_{|J|=\ell,|K|=m} b_{J,K} dz_J \wedge d\bar{z}_K \right)}_{\in \Omega^{\ell,m}},
$$

$$d\beta = \sum_{\ell+m=k} \left(\sum_{|J|=\ell,|K|=m} db_{J,K} \wedge dz_J \wedge d\bar{z}_K \right)$$

$$= \sum_{\ell+m=k} \sum_{|J|=\ell,|K|=m} \left(\partial b_{J,K} + \bar{\partial} b_{J,K} \right) \wedge dz_J \wedge d\bar{z}_K$$

(because $d = \partial + \bar{\partial}$ on functions)

$$= \sum_{\ell+m=k} \left(\underbrace{\sum_{|J|=\ell,|K|=m} \partial b_{J,K} \wedge dz_J \wedge d\bar{z}_K}_{\in \Omega^{\ell+1,m}} + \underbrace{\sum_{|J|=\ell,|K|=m} \bar{\partial} b_{J,K} \wedge dz_J \wedge d\bar{z}_K}_{\in \Omega^{\ell,m+1}} \right)$$

$$= \partial\beta + \bar{\partial}\beta .$$

Therefore, $d = \partial + \bar{\partial}$ on forms of any degree for a *complex* manifold.

Conclusion. If M is a complex manifold, then $d = \partial + \bar{\partial}$. (For an almost complex manifold this fails because there are no coordinate functions z_j to give a suitable basis of 1-forms.)

Remark. If $b \in C^\infty(\mathcal{U};\mathbb{C})$, in terms of z and $\bar{z}$, we obtain the following formulas:

$$db = \sum_j \left(\frac{\partial b}{\partial x_j} dx_j + \frac{\partial b}{\partial y_j} dy_j \right)$$

$$= \sum_j \left[\frac{1}{2}\left(\frac{\partial b}{\partial x_j} - i\frac{\partial b}{\partial y_j} \right)(dx_j + idy_j) + \frac{1}{2}\left(\frac{\partial b}{\partial x_j} + i\frac{\partial b}{\partial y_j} \right)(dx_j - idy_j) \right]$$

$$= \sum_j \left(\frac{\partial b}{\partial z_j} dz_j + \frac{\partial b}{\partial \bar{z}_j} d\bar{z}_j \right).$$

Hence:

$$\begin{cases} \partial b = \pi^{1,0} db = \sum_j \frac{\partial b}{\partial z_j} dz_j \\ \bar{\partial} b = \pi^{0,1} db = \sum_j \frac{\partial b}{\partial \bar{z}_j} d\bar{z}_j \end{cases}$$

$\Diamond$

In the case where $\beta \in \Omega^{\ell,m}$, we have

$$d\beta = \partial\beta + \bar{\partial}\beta = (\ell+1,m)\text{-form} + (\ell,m+1)\text{-form}$$
$$0 = d^2\beta = (\ell+2,m)\text{-form} + (\ell+1,m+1)\text{-form} + (\ell,m+2)\text{-form}$$

$$= \underbrace{\partial^2\beta}_{0} + \underbrace{(\partial\bar{\partial} + \bar{\partial}\partial)\beta}_{0} + \underbrace{\bar{\partial}^2\beta}_{0} .$$

Hence, $\bar{\partial}^2 = 0$.

The Dolbeault theorem states that for complex manifolds

$$H_{\text{Dolbeault}}^{\ell,m}(M) = H^m(M; \mathcal{O}(\Omega^{(\ell,0)})) \,,$$

where $\mathcal{O}(\Omega^{(\ell,0)})$ is the sheaf of forms of type $(\ell,0)$ over M.

Theorem 15.4. *(Newlander-Nirenberg, 1957 [89])*
    *Let (M,J) be an almost complex manifold. Let $\mathcal{N}$ be the **Nijenhuis tensor** (defined in Homework 10). Then:*

$$
\begin{aligned}
M \text{ is a complex manifold} &\iff J \text{ is integrable} \\
&\iff \mathcal{N} \equiv 0 \\
&\iff d = \partial + \bar{\partial} \\
&\iff \bar{\partial}^2 = 0 \\
&\iff \pi^{2,0} d|_{\Omega^{0,1}} = 0 \,.
\end{aligned}
$$

For the proof of this theorem, besides the original reference, see also [22, 30, 48, 66, 109]. Naturally most almost complex manifolds have $d \neq \partial + \bar{\partial}$.

Homework 11: Complex Projective Space

The complex projective space $\mathbb{CP}^n$ is the space of complex lines in $\mathbb{C}^{n+1}$:

$\mathbb{CP}^n$ is obtained from $\mathbb{C}^{n+1} \setminus \{0\}$ by making the identifications $(z_0, \ldots, z_n) \sim (\lambda z_0, \ldots, \lambda z_n)$ for all $\lambda \in \mathbb{C} \setminus \{0\}$. One denotes by $[z_0, \ldots, z_n]$ the equivalence class of $(z_0, \ldots, z_n)$, and calls $z_0, \ldots, z_n$ the homogeneous coordinates of the point $p = [z_0, \ldots, z_n]$. (The homogeneous coordinates are, of course, only determined up to multiplication by a non-zero complex number λ.)

Let $\mathcal{U}_i$ be the subset of $\mathbb{CP}^n$ consisting of all points $p = [z_0, \ldots, z_n]$ for which $z_i \neq 0$. Let $\varphi_i : \mathcal{U}_i \to \mathbb{C}^n$ be the map

$$\varphi_i([z_0, \ldots, z_n]) = \left(\frac{z_0}{z_i}, \ldots, \frac{z_{i-1}}{z_i}, \frac{z_{i+1}}{z_i}, \ldots, \frac{z_n}{z_i} \right).$$

1. Show that the collection

$$\{(\mathcal{U}_i, \mathbb{C}^n, \varphi_i), i = 0, \ldots, n\}$$

   is an atlas in the *complex* sense, i.e., the transition maps are biholomorphic. Conclude that $\mathbb{CP}^n$ is a complex manifold.

   Hint: Work out the transition maps associated with $(\mathcal{U}_0, \mathbb{C}^n, \varphi_0)$ and $(\mathcal{U}_1, \mathbb{C}^n, \varphi_1)$. Show that the transition diagram has the form

   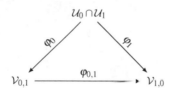

   where $\mathcal{V}_{0,1} = \mathcal{V}_{1,0} = \{(z_1, \ldots, z_n) \in \mathbb{C}^n \,|\, z_1 \neq 0\}$ and

   $$\varphi_{0,1}(z_1, \ldots, z_n) = \left(\frac{1}{z_1}, \frac{z_2}{z_1}, \ldots, \frac{z_n}{z_1} \right).$$

2. Show that the 1-dimensional complex manifold $\mathbb{CP}^1$ is diffeomorphic, as a real 2-dimensional manifold, to S^2.

   Hint: Stereographic projection.

Chapter 16
Kähler Forms

16.1 Kähler Forms

Definition 16.1. A *Kähler manifold* is a symplectic manifold (M, ω) equipped with an integrable compatible almost complex structure. The symplectic form ω is then called a *Kähler form*.

It follows immediately from the previous definition that

$$(M, \omega) \text{ is Kähler} \implies M \text{ is a complex manifold}$$

$$\implies \begin{cases} \Omega^k(M; \mathbb{C}) = \oplus_{\ell+m=k} \Omega^{\ell,m} \\ d = \partial + \bar{\partial} \end{cases}$$

where

$$\partial = \pi^{\ell+1,m} \circ d : \Omega^{\ell,m} \to \Omega^{\ell+1,m}$$
$$\bar{\partial} = \pi^{\ell,m+1} \circ d : \Omega^{\ell,m} \to \Omega^{\ell,m+1} .$$

On a complex chart $(\mathcal{U}, z_1, \ldots, z_n)$, $n = \dim_{\mathbb{C}} M$,

$$\Omega^{\ell,m} = \left\{ \sum_{|J|=\ell, |K|=m} b_{JK} dz_J \wedge d\bar{z}_K \mid b_{JK} \in C^\infty(\mathcal{U}; \mathbb{C}) \right\} ,$$

where

$$J = (j_1, \ldots, j_\ell) , \quad j_1 < \ldots < j_\ell , \quad dz_J = dz_{j_1} \wedge \ldots \wedge dz_{j_\ell} ,$$
$$K = (k_1, \ldots, k_m) , \quad k_1 < \ldots < k_m , \quad d\bar{z}_K = d\bar{z}_{k_1} \wedge \ldots \wedge d\bar{z}_{k_m} .$$

On the other hand,

$$(M, \omega) \text{ is Kähler} \implies \omega \text{ is a symplectic form} .$$

– Where does ω fit with respect to the above decomposition?

A Kähler form ω is

1. a 2-form,
2. compatible with the complex structure,
3. closed,
4. real-valued, and
5. nondegenerate.

These properties translate into:

1. $\Omega^2(M;\mathbb{C}) = \Omega^{2,0} \oplus \Omega^{1,1} \oplus \Omega^{0,2}$.
   On a local complex chart $(\mathcal{U}, z_1, \ldots, z_n)$,

$$\omega = \sum a_{jk}\, dz_j \wedge dz_k + \sum b_{jk}\, dz_j \wedge d\bar{z}_k + \sum c_{jk}\, d\bar{z}_j \wedge d\bar{z}_k$$

   for some $a_{jk}, b_{jk}, c_{jk} \in C^\infty(\mathcal{U};\mathbb{C})$.
2. J is a symplectomorphism, that is, $J^*\omega = \omega$ where $(J^*\omega)(u,v) := \omega(Ju, Jv)$.

$$J^* dz_j = dz_j \circ J = \ idz_j$$
$$J^* d\bar{z}_j = d\bar{z}_j \circ J = -id\bar{z}_j$$

$$J^*\omega = \sum \overset{\overset{-1}{\|}}{(i \cdot i)}\, a_{jk}\, dz_j \wedge dz_k + \overset{\overset{1}{\|}}{i(-i)} \sum b_{jk}\, dz_j \wedge d\bar{z}_k + \overset{\overset{-1}{\|}}{(-i)^2} \sum c_{jk} d\bar{z}_j \wedge d\bar{z}_k$$

$$J^*\omega = \omega \quad \Longleftrightarrow \quad a_{jk} = 0 = c_{jk}, \text{ all } j,k \quad \Longleftrightarrow \quad \omega \in \Omega^{1,1}.$$

3. $0 = d\omega = \underbrace{\partial\omega}_{(2,1)-\text{form}} + \underbrace{\bar{\partial}\omega}_{(1,2)-\text{form}} \Longrightarrow \begin{cases} \partial\omega = 0 & \omega \text{ is } \partial\text{-closed} \\ \bar{\partial}\omega = 0 & \omega \text{ is } \bar{\partial}\text{-closed} \end{cases}$

   Hence, ω defines a Dolbeault $(1,1)$ cohomology class,

$$[\omega] \in H^{1,1}_{\text{Dolbeault}}(M).$$

   Putting $b_{jk} = \frac{i}{2} h_{jk}$,

$$\omega = \frac{i}{2} \sum_{j,k=1}^{n} h_{jk}\, dz_j \wedge d\bar{z}_k, \qquad h_{jk} \in C^\infty(\mathcal{U};\mathbb{C}).$$

4. ω real-valued $\Longleftrightarrow \omega = \bar{\omega}$.

$$\bar{\omega} = -\frac{i}{2} \sum \overline{h_{jk}}\, d\bar{z}_j \wedge dz_k = \frac{i}{2} \sum \overline{h_{jk}}\, dz_k \wedge d\bar{z}_j = \frac{i}{2} \sum \overline{h_{kj}}\, dz_j \wedge d\bar{z}_k$$

$$\omega \text{ real} \quad \Longleftrightarrow \quad h_{jk} = \overline{h_{kj}},$$

   i.e., at every point $p \in \mathcal{U}$, the $n \times n$ matrix $(h_{jk}(p))$ is hermitian.
5. nondegeneracy: $\omega^n = \underbrace{\omega \wedge \ldots \wedge \omega}_{n} \neq 0$.

Exercise. Check that

$$\omega^n = n! \left(\frac{i}{2}\right)^n \det(h_{jk}) \, dz_1 \wedge d\bar{z}_1 \wedge \ldots \wedge dz_n \wedge d\bar{z}_n .$$

$\diamond$

Now

$$\omega \text{ nondegenerate} \iff \det_{\mathbb{C}}(h_{jk}) \neq 0 ,$$

i.e., at every $p \in M$, $(h_{jk}(p))$ is a nonsingular matrix.
2. Again the positivity condition: $\omega(v, Jv) > 0$, $\forall v \neq 0$.

Exercise. Show that $(h_{jk}(p))$ is positive-definite. $\diamond$

$$\omega \text{ positive} \iff (h_{jk}) \gg 0 ,$$

i.e., at each $p \in \mathcal{U}$, $(h_{jk}(p))$ is positive-definite.

Conclusion. Kähler forms are ∂- and $\bar{\partial}$-closed $(1,1)$-forms, which are given on a local chart $(\mathcal{U}, z_1, \ldots, z_n)$ by

$$\omega = \frac{i}{2} \sum_{j,k=1}^n h_{jk} \, dz_j \wedge d\bar{z}_k$$

where, at every point $p \in \mathcal{U}$, $(h_{jk}(p))$ is a positive-definite hermitian matrix.

16.2 An Application

Theorem 16.2. (Banyaga) *Let M be a compact complex manifold. Let ω_0 and ω_1 be Kähler forms on M. If $[\omega_0] = [\omega_1] \in H^2_{\mathrm{deRham}}(M)$, then (M, ω_0) and (M, ω_1) are symplectomorphic.*

Proof. Any combination $\omega_t = (1-t)\omega_0 + t\omega_1$ is symplectic for $0 \leq t \leq 1$, because, on a complex chart $(\mathcal{U}, z_1, \ldots, z_n)$, where $n = \dim_{\mathbb{C}} M$, we have

$$\omega_0 = \tfrac{i}{2} \sum h^0_{jk} dz_j \wedge d\bar{z}_k$$

$$\omega_1 = \tfrac{i}{2} \sum h^1_{jk} dz_j \wedge dz_k$$

$$\omega_t = \tfrac{i}{2} \sum h^t_{jk} dz_j \wedge d\bar{z}_k , \quad \text{where } h^t_{jk} = (1-t)h^0_{jk} + th^1_{jk} .$$

$$(h^0_{jk}) \gg 0 , (h^1_{jk}) \gg 0 \implies (h^t_{jk}) \gg 0 .$$

Apply the Moser theorem (Theorem 7.2). $\square$

16.3 Recipe to Obtain Kähler Forms

Definition 16.3. Let M be a complex manifold. A function $\rho \in C^\infty(M; \mathbb{R})$ is **strictly plurisubharmonic** (s.p.s.h.) if, on each local complex chart $(\mathcal{U}, z_1, \dots, z_n)$, where $n = \dim_\mathbb{C} M$, the matrix $\left(\frac{\partial^2 \rho}{\partial z_j \partial \bar{z}_k}(p) \right)$ is positive-definite at all $p \in \mathcal{U}$.

Proposition 16.4. *Let M be a complex manifold and let $\rho \in C^\infty(M; \mathbb{R})$ be s.p.s.h.. Then*

$$\omega = \frac{i}{2} \partial \bar{\partial} \rho \qquad \text{is Kähler} \,.$$

A function ρ as in the previous proposition is called a (global) **Kähler potential**.

Proof. Simply observe that:

$$
\begin{cases}
\partial \omega = \frac{i}{2} \overbrace{\partial^2}^{0} \bar{\partial} \rho = 0 \\[2ex]
\bar{\partial} \omega = \frac{i}{2} \underbrace{\bar{\partial} \partial}_{-\partial \bar{\partial}} \bar{\partial} \rho = -\frac{i}{2} \partial \underbrace{\bar{\partial}^2}_{0} \rho = 0
\end{cases}
$$

$$d\omega = \partial \omega + \bar{\partial} \omega = 0 \implies \omega \text{ is closed} \,.$$

$$\bar{\omega} = -\frac{i}{2} \bar{\partial} \partial \rho = \frac{i}{2} \partial \bar{\partial} \rho = \omega \implies \omega \text{ is real} \,.$$

$$\omega \in \Omega^{1,1} \implies J^* \omega = \omega \implies \omega(\cdot, J\cdot) \text{ is symmetric} \,.$$

Exercise. Show that, for $f \in C^\infty(\mathcal{U}; \mathbb{C})$,

$$\partial f = \sum \frac{\partial f}{\partial z_j} dz_j \quad \text{and} \quad \bar{\partial} f = \sum \frac{\partial f}{\partial \bar{z}_j} d\bar{z}_j \,.$$

Since the right-hand sides are in $\Omega^{1,0}$ and $\Omega^{0,1}$, respectively, it suffices to show that the sum of the two expressions is df. $\diamondsuit$

$$\omega = \frac{i}{2} \partial \bar{\partial} \rho = \frac{i}{2} \sum \frac{\partial}{\partial z_j} \left(\frac{\partial \rho}{\partial \bar{z}_k} \right) dz_j \wedge d\bar{z}_k = \frac{i}{2} \sum \underbrace{\left(\frac{\partial^2 \rho}{\partial z_j \partial \bar{z}_k} \right)}_{h_{jk}} dz_j \wedge d\bar{z}_k \,.$$

$$\rho \text{ is s.p.s.h} \implies (h_{jk}) \gg 0 \implies \omega(\cdot, J\cdot) \text{ is positive} \,.$$

In particular, ω is nondegenerate. $\square$

Example. Let $M = \mathbb{C}^n \simeq \mathbb{R}^{2n}$, with complex coordinates $(z_1, \dots, z_n)$ and corresponding real coordinates $(x_1, y_1, \dots, x_n, y_n)$ via $z_j = x_j + iy_j$. Let

$$\rho(x_1, y_1, \ldots, x_n, y_n) = \sum_{j=1}^{n} (x_j^2 + y_j^2) = \sum |z_j|^2 = \sum z_j \bar{z}_j .$$

Then

$$\frac{\partial}{\partial z_j} \frac{\partial \rho}{\partial \bar{z}_k} = \frac{\partial}{\partial z_j} z_k = \delta_{jk} ,$$

so

$$(h_{jk}) = \left(\frac{\partial^2 \rho}{\partial z_j \partial \bar{z}_k} \right) = (\delta_{jk}) = \mathrm{Id} \gg 0 \implies \rho \text{ is s.p.s.h. .}$$

The corresponding Kähler form

$$\omega = \frac{i}{2} \partial \bar{\partial} \rho = \frac{i}{2} \sum_{j,k} \delta_{jk} \, dz_j \wedge d\bar{z}_k$$

$$= \frac{i}{2} \sum_j dz_j \wedge d\bar{z}_j = \sum_j dx_j \wedge dy_j \quad \text{is the standard form .}$$

16.4 Local Canonical Form for Kähler Forms

There is a local converse to the previous construction of Kähler forms.

Theorem 16.5. *Let ω be a closed real-valued $(1,1)$-form on a complex manifold M and let $p \in M$. Then there exist a neighborhood $\mathcal{U}$ of p and $\rho \in C^\infty(\mathcal{U}; \mathbb{R})$ such that, on $\mathcal{U}$,*

$$\omega = \frac{i}{2} \partial \bar{\partial} \rho .$$

The function ρ is then called a (local) **Kähler potential**.

The proof requires holomorphic versions of Poincaré's lemma, namely, the local triviality of Dolbeault groups:

$$\forall p \in M \quad \exists \text{ neighborhood } \mathcal{U} \text{ of } p \text{ such that } \quad H_{\text{Dolbeault}}^{\ell,m}(\mathcal{U}) = 0 , \, m > 0 ,$$

and the local triviality of the holomorphic de Rham groups; see [48].

Proposition 16.6. *Let M be a complex manifold, $\rho \in C^\infty(M; \mathbb{R})$ s.p.s.h., X a complex submanifold, and $i : X \hookrightarrow M$ the inclusion map. Then $i^*\rho$ is s.p.s.h..*

Proof. Let $\dim_\mathbb{C} M = n$ and $\dim_\mathbb{C} X = n - m$. For $p \in X$, choose a chart $(\mathcal{U}, z_1, \ldots, z_n)$ for M centered at p and adapted to X, i.e., $X \cap \mathcal{U}$ is given by $z_1 = \ldots = z_m = 0$. In this chart, $i^*\rho = \rho(0, 0, \ldots, 0, z_{m+1}, \ldots, z_n)$.

$$i^* p \text{ is s.p.s.h.} \iff \left(\frac{\partial^2 \rho}{\partial z_{m+j} \partial \bar{z}_{m+k}}(0,\ldots,0,z_{m+1},\ldots,z_n) \right) \text{ is positive-definite} ,$$

which holds since this is a minor of $\left(\frac{\partial^2}{\partial z_j \partial \bar{z}_k}(0,\ldots,0,z_{m+1},\ldots,z_n) \right)$. $\qquad \square$

Corollary 16.7. *Any complex submanifold of a Kähler manifold is also Kähler.*

Definition 16.8. Let (M,ω) be a Kähler manifold, X a complex submanifold, and $i : X \hookrightarrow M$ the inclusion. Then $(X, i^*\omega)$ is called a ***Kähler submanifold***.

Example. Complex vector space $(\mathbb{C}^n, \omega_0)$ where $\omega_0 = \frac{i}{2}\sum dz_j \wedge d\bar{z}_j$ is Kähler. Every complex submanifold of $\mathbb{C}^n$ is Kähler. $\qquad \diamond$

Example. The complex projective space is

$$\mathbb{CP}^n = \mathbb{C}^{n+1}\setminus\{0\}/ \sim$$

where

$$(z_0,\ldots,z_n) \sim (\lambda z_0,\ldots,\lambda z_n) , \quad \lambda \in \mathbb{C}\setminus\{0\} .$$

The Fubini-Study form (see Homework 12) is Kähler. Therefore, every **non-singular projective variety** is a Kähler submanifold. Here we mean

$$
\begin{aligned}
\text{non-singular} &= \text{smooth} \\
\text{projective variety} &= \text{zero locus of a collection} \\
& \text{of homogeneous polynomials} .
\end{aligned}
$$

Homework 12: The Fubini-Study Structure

The purpose of the following exercises is to describe the natural Kähler structure on complex projective space, $\mathbb{CP}^n$.

1. Show that the function on $\mathbb{C}^n$

$$z \longmapsto \log(|z|^2 + 1)$$

is strictly plurisubharmonic. Conclude that the 2-form

$$\omega_{FS} = \tfrac{i}{2} \partial \bar{\partial} \log(|z|^2 + 1)$$

is a Kähler form. (It is usually called the **Fubini-Study form** on $\mathbb{C}^n$.)

> **Hint:** A hermitian $n \times n$ matrix H is positive definite if and only if $v^* H v > 0$ for any $v \in \mathbb{C}^n \setminus \{0\}$, where v^* is the transpose of the vector $\bar{v}$. To prove positive-definiteness, either apply the Cauchy-Schwarz inequality, or use the following symmetry observation: $U(n)$ acts transitively on S^{2n-1} and ω_{FS} is $U(n)$-invariant, thus it suffices to show positive-definiteness along *one* direction.

2. Let $\mathcal{U}$ be the open subset of $\mathbb{C}^n$ defined by the inequality $z_1 \neq 0$, and let $\varphi : \mathcal{U} \to \mathcal{U}$ be the map

$$\varphi(z_1, \ldots, z_n) = \tfrac{1}{z_1}(1, z_2, \ldots, z_n) \ .$$

Show that φ maps $\mathcal{U}$ biholomorphically onto $\mathcal{U}$ and that

$$\varphi^* \log(|z|^2 + 1) = \log(|z|^2 + 1) + \log \tfrac{1}{|z_1|^2} \ . \qquad (\star)$$

3. Notice that, for every point $p \in \mathcal{U}$, we can write the second term in $(\star)$ as the sum of a holomorphic and an anti-holomorphic function:

$$-\log z_1 - \log \overline{z_1}$$

on a neighborhood of p. Conclude that

$$\partial \bar{\partial} \varphi^* \log(|z|^2 + 1) = \partial \bar{\partial} \log(|z|^2 + 1)$$

and hence that $\varphi^* \omega_{FS} = \omega_{FS}$.

> **Hint:** You need to use the fact that the pullback by a holomorphic map φ^* commutes with the ∂ and $\bar{\partial}$ operators. This is a consequence of φ^* preserving form type, $\varphi^*(\Omega^{p,q}) \subseteq \Omega^{p,q}$, which in turn is implied by $\varphi^* dz_j = \partial \varphi_j \subset \Omega^{1,0}$ and $\varphi^* d\overline{z_j} = \bar{\partial} \overline{\varphi_j} \subseteq \Omega^{0,1}$, where φ_j is the jth component of φ with respect to local complex coordinates $(z_1, \ldots, z_n)$.

4. Recall that $\mathbb{CP}^n$ is obtained from $\mathbb{C}^{n+1} \setminus \{0\}$ by making the identifications $(z_0, \ldots, z_n) \sim (\lambda z_0, \ldots, \lambda z_n)$ for all $\lambda \in \mathbb{C} \setminus \{0\}$; $[z_0, \ldots, z_n]$ is the equivalence class of $(z_0, \ldots, z_n)$.

For $i = 0, 1, \ldots, n$, let

$$\mathcal{U}_i = \{[z_0, \ldots, z_n] \in \mathbb{CP}^n \,|\, z_i \neq 0\}$$
$$\varphi_i : \mathcal{U}_i \to \mathbb{C}^n \qquad \varphi_i([z_0, \ldots, z_n]) = \left(\tfrac{z_0}{z_i}, \ldots, \tfrac{z_{i-1}}{z_i}, \tfrac{z_{i+1}}{z_i}, \ldots, \tfrac{z_n}{z_i} \right).$$

Homework 11 showed that the collection $\{(\mathcal{U}_i, \mathbb{C}^n, \varphi_i), i = 0, \ldots, n\}$ is a complex atlas (i.e., the transition maps are biholomorphic). In particular, it was shown that the transition diagram associated with $(\mathcal{U}_0, \mathbb{C}^n, \varphi_0)$ and $(\mathcal{U}_1, \mathbb{C}^n, \varphi_1)$ has the form

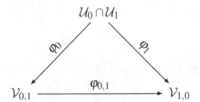

where $\mathcal{V}_{0,1} = \mathcal{V}_{1,0} = \{(z_1, \ldots, z_n) \in \mathbb{C}^n \,|\, z_1 \neq 0\}$ and $\varphi_{0,1}(z_1, \ldots, z_n) = (\tfrac{1}{z_1}, \tfrac{z_2}{z_1}, \ldots, \tfrac{z_n}{z_1})$.
Now the set $\mathcal{U}$ in exercise 2 is equal to the sets $\mathcal{V}_{0,1}$ and $\mathcal{V}_{1,0}$, and the map φ coincides with $\varphi_{0,1}$.

Show that $\varphi_0^* \omega_{FS}$ and $\varphi_1^* \omega_{FS}$ are identical on the overlap $\mathcal{U}_0 \cap \mathcal{U}_1$.

More generally, show that the Kähler forms $\varphi_i^* \omega_{FS}$ "glue together" to define a Kähler structure on $\mathbb{CP}^n$. This is called the **Fubini-Study form** on complex projective space.

5. Prove that for $\mathbb{CP}^1$ the Fubini-Study form on the chart $\mathcal{U}_0 = \{[z_0, z_1] \in \mathbb{CP}^1 \,|\, z_0 \neq 0\}$ is given by the formula

$$\omega_{FS} = \frac{dx \wedge dy}{(x^2 + y^2 + 1)^2}$$

where $\tfrac{z_1}{z_0} = z = x + iy$ is the usual coordinate on $\mathbb{C}$.

6. Compute the total area of $\mathbb{CP}^1 = \mathbb{C} \cup \{\infty\}$ with respect to ω_{FS}:

$$\int_{\mathbb{CP}^1} \omega_{FS} = \int_{\mathbb{R}^2} \frac{dx \wedge dy}{(x^2 + y^2 + 1)^2}.$$

7. Recall that $\mathbb{CP}^1 \simeq S^2$ as real 2-dimensional manifolds (Homework 11). On S^2 there is the standard area form ω_{std} induced by regarding it as the unit sphere in $\mathbb{R}^3$ (Homework 6): in cylindrical polar coordinates (θ, h) on S^2 away from its poles ($0 \leq \theta < 2\pi$ and $-1 \leq h \leq 1$), we have

$$\omega_{std} = d\theta \wedge dh.$$

Using stereographic projection, show that

$$\omega_{FS} = \frac{1}{4} \omega_{std}.$$

Chapter 17
Compact Kähler Manifolds

17.1 Hodge Theory

Let M be a complex manifold. A Kähler form ω on M is a symplectic form which is compatible with the complex structure. Equivalently, a Kähler form ω is a ∂- and $\bar{\partial}$-closed form of type $(1,1)$ which, on a local chart $(\mathcal{U}, z_1, \ldots, z_n)$ is given by $\omega = \frac{i}{2} \sum_{j,k=1}^{n} h_{jk} dz_j \wedge d\bar{z}_k$, where, at each $x \in \mathcal{U}$, $(h_{jk}(x))$ is a positive-definite hermitian matrix. The pair (M, ω) is then called a Kähler manifold.

Theorem 17.1. (Hodge) *On a compact Kähler manifold (M, ω) the Dolbeault cohomology groups satisfy*

$$H^k_{\text{deRham}}(M; \mathbb{C}) \simeq \bigoplus_{\ell+m=k} H^{\ell,m}_{\text{Dolbeault}}(M) \qquad \textbf{\textit{(Hodge decomposition)}}$$

with $H^{\ell,m} \simeq \overline{H^{m,\ell}}$. In particular, the spaces $H^{\ell,m}_{\text{Dolbeault}}$ are finite-dimensional.

Hodge identified the spaces of cohomology classes of forms with spaces of actual forms, by picking *the* representative from each class which solves a certain differential equation, namely the *harmonic* representative.

(1) The **Hodge $*$-operator**.

Each tangent space $V = T_x M$ has a positive inner product $\langle \cdot, \cdot \rangle$, part of the riemannian metric in a compatible triple; we forget about the complex and symplectic structures until part (4).

Let $e_1, \ldots, e_n$ be a positively oriented orthonormal basis of V
The star operator is a linear operator $* : \Lambda(V) \to \Lambda(V)$ defined by

$$*(1) = e_1 \wedge \ldots \wedge e_n$$
$$*(e_1 \wedge \ldots \wedge e_n) = 1$$
$$*(e_1 \wedge \ldots \wedge e_k) = e_{k+1} \wedge \ldots \wedge e_n .$$

We see that $* : \Lambda^k(V) \to \Lambda^{n-k}(V)$ and satisfies $** = (-1)^{k(n-k)}$.

(2) The **codifferential** and the **laplacian** are the operators defined by:

$$\delta = (-1)^{n(k+1)+1} * d * \; : \Omega^k(M) \to \Omega^{k-1}(M)$$
$$\Delta = d\delta + \delta d \qquad \qquad : \Omega^k(M) \to \Omega^k(M) \; .$$

The operator Δ is also called the **Laplace-Beltrami operator**.

Exercise. Check that, on $\Omega^0(\mathbb{R}^n) = C^\infty(\mathbb{R}^n)$, $\Delta = -\sum_{i=1}^n \frac{\partial^2}{\partial x_i^2}$. $\Diamond$

Exercise. Check that $\Delta* = *\Delta$. $\Diamond$

Suppose that M is compact. Define an inner product on forms by

$$\langle \cdot, \cdot \rangle : \Omega^k \times \Omega^k \to \mathbb{R} \; , \qquad \langle \alpha, \beta \rangle = \int_M \alpha \wedge *\beta \; .$$

Exercise. Check that this is symmetric, positive-definite and satisfies $\langle d\alpha, \beta \rangle = \langle \alpha, \delta\beta \rangle$. $\Diamond$

Therefore, δ is often denoted by d^* and called the adjoint of d. (When M is not compact, we still have a formal adjoint of d with respect to the nondegenerate bilinear pairing $\langle \cdot, \cdot \rangle : \Omega^k \times \Omega^k_c \to \mathbb{R}$ defined by a similar formula, where Ω^k_c is the space of compactly supported k-forms.) Also, Δ is self-adjoint:

Exercise. Check that $\langle \Delta\alpha, \beta \rangle = \langle \alpha, \Delta\beta \rangle$, and that $\langle \Delta\alpha, \alpha \rangle = |d\alpha|^2 + |\delta\alpha|^2 \geq 0$, where $|\cdot|$ is the norm with respect to this inner product. $\Diamond$

(3) The **harmonic k-forms** are the elements of $\mathcal{H}^k := \{\alpha \in \Omega^k \mid \Delta\alpha = 0\}$.

Note that $\Delta\alpha = 0 \iff d\alpha = \delta\alpha = 0$. Since a harmonic form is d-closed, it defines a de Rham cohomology class.

Theorem 17.2. (Hodge) *Every de Rham cohomology class on a compact oriented riemannian manifold (M, g) possesses a unique harmonic representative, i.e.,*

$$\mathcal{H}^k \simeq H^k_{\text{deRham}}(M; \mathbb{R}) \; .$$

In particular, the spaces $\mathcal{H}^k$ are finite-dimensional. We also have the following orthogonal decomposition with respect to $\langle \cdot, \cdot \rangle$:

$$\Omega^k \simeq \mathcal{H}^k \oplus \Delta(\Omega^k(M))$$
$$\simeq \mathcal{H}^k \oplus d\Omega^{k-1} \oplus \delta\Omega^{k+1} \qquad \textbf{(Hodge decomposition on forms)} \; .$$

The proof involves functional analysis, elliptic differential operators, pseudodifferential operators and Fourier analysis; see [48, 109].

So far, this was ordinary Hodge theory, considering only the metric and not the complex structure.

(4) **Complex Hodge Theory.**
When M is Kähler, the laplacian satisfies $\Delta = 2(\bar{\partial}\bar{\partial}^* + \bar{\partial}^*\bar{\partial})$ (see, for example, [48])and preserves the decomposition according to type, $\Delta : \Omega^{\ell,m} \to \Omega^{\ell,m}$. Hence, harmonic forms are also bigraded

$$\mathcal{H}^k = \bigoplus_{\ell+m=k} \mathcal{H}^{\ell,m} .$$

Theorem 17.3. (Hodge) *Every Dolbeault cohomology class on a compact Kähler manifold (M,ω) possesses a unique harmonic representative, i.e.,*

$$\mathcal{H}^{\ell,m} \simeq H^{\ell,m}_{\text{Dolbeault}}(M)$$

and the spaces $\mathcal{H}^{\ell,m}$ are finite-dimensional. Hence, we have the following isomorphisms:

$$H^k_{\text{deRham}}(M) \simeq \mathcal{H}^k \simeq \bigoplus_{\ell+m=k} \mathcal{H}^{\ell,m} \simeq \bigoplus_{\ell+m=k} H^{\ell,m}_{\text{Dolbeault}}(M) .$$

For the proof, see for instance [48, 109].

17.2 Immediate Topological Consequences

Let $b^k(M) := \dim H^k_{\text{deRham}}(M)$ be the usual **Betti numbers** of M, and let $h^{\ell,m}(M) := \dim H^{\ell,m}_{\text{Dolbeault}}(M)$ be the so-called **Hodge numbers** of M.

$$\text{Hodge Theorem} \quad \Longrightarrow \quad \begin{cases} b^k = \sum_{\ell+m=k} h^{\ell,m} \\ h^{\ell,m} = h^{m,\ell} \end{cases}$$

Some immediate topological consequences are:

1. On compact Kähler manifolds "the odd Betti numbers are even:"

$$b^{2k+1} = \sum_{\ell+m=2k+1} h^{\ell,m} = 2\sum_{\ell=0}^{k} h^{\ell,(2k+1-\ell)} \quad \text{is even} .$$

2. On compact Kähler manifolds, $h^{1,0} = \frac{1}{2}b^1$ is a topological invariant.
3. On compact symplectic manifolds, "even Betti numbers are positive," because ω^k is closed but not exact $(k = 0, 1, \ldots, n)$.

   Proof. If $\omega^k = d\alpha$, by Stokes' theorem, $\int_M \omega^n = \int_M d(\alpha \wedge \omega^{n-k}) = 0$.
   This cannot happen since ω^n is a volume form. $\qquad\qquad\qquad\square$

4. On compact Kähler manifolds, the $h^{\ell,\ell}$ are positive.

Claim. $0 \neq [\omega^\ell] \in H^{\ell,\ell}_{\text{Dolbeault}}(M)$.

Proof.

$$\omega \in \Omega^{1,1} \implies \omega^\ell \in \Omega^{\ell,\ell}$$
$$d\omega = 0 \implies 0 = d\omega^\ell = \underbrace{\partial \omega^\ell}_{(\ell+1,\ell)} + \underbrace{\bar{\partial} \omega^\ell}_{(\ell,\ell+1)}$$
$$\implies \bar{\partial} \omega^\ell = 0 \,,$$

so $[\omega^\ell]$ defines an element of $H^{\ell,\ell}_{\text{Dolbeault}}$. Why is ω^ℓ not $\bar{\partial}$-exact?
If $\omega^\ell = \bar{\partial}\beta$ for some $\beta \in \Omega^{\ell-1,\ell}$, then

$$\omega^n = \omega^\ell \wedge \omega^{n-\ell} = \bar{\partial}(\beta \wedge \omega^{n-\ell}) \implies 0 = [\omega^n] \in H^{n,n}_{\text{Dolbeault}}(M) \,.$$

But $[\omega^n] \neq 0$ in $H^{2n}_{\text{deRham}}(M;\mathbb{C}) \simeq H^{n,n}_{\text{Dolbeault}}(M)$ since it is a volume form.

$\square$

There are other constraints on the Hodge numbers of compact Kähler manifolds, and ongoing research on how to compute $H^{\ell,m}_{\text{Dolbeault}}$. A popular picture to describe the relations is the **Hodge diamond**:

$$
\begin{array}{ccccccccc}
 & & & & h^{n,n} & & & & \\
 & & & h^{n,n-1} & & h^{n-1,n} & & & \\
 & & h^{n,n-2} & & h^{n-1,n-1} & & h^{n-2,n} & & \\
 & & & & \vdots & & & & \\
 & h^{2,0} & & & h^{1,1} & & & h^{0,2} & \\
 & & h^{1,0} & & & h^{0,1} & & & \\
 & & & & h^{0,0} & & & &
\end{array}
$$

Complex conjugation gives symmetry with respect to the middle vertical, whereas the Hodge $*$ induces symmetry about the center of the diamond. The middle vertical axis is all non-zero. There are further symmetries induced by isomorphisms given by wedging with ω.

The **Hodge conjecture** relates $H^{\ell,\ell}_{\text{Dolbeault}}(M) \cap H^{2\ell}(M;\mathbb{Z})$ for projective manifolds M (i.e., submanifolds of complex projective space) to $\text{codim}_\mathbb{C} = \ell$ complex submanifolds of M.

17.3 Compact Examples and Counterexamples

$$
\begin{array}{ccc}
\text{symplectic} & \Longleftarrow & \text{Kähler} \\
\Downarrow & & \Downarrow \\
\text{almost complex} & \Longleftarrow & \text{complex}
\end{array}
$$

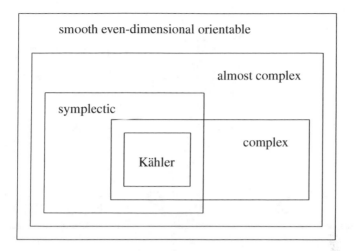

Is each of these regions nonempty? Can we even find representatives of each region which are simply connected or have any specified fundamental group?

- Not all smooth even-dimensional manifolds are almost complex. For example, S^4, S^8, S^{10}, etc., are not almost complex.
- If M is both symplectic and complex, is it necessarily Kähler?

No. For some time, it had been suspected that every compact symplectic manifold might have an underlying Kähler structure, or, at least, that a compact symplectic manifold might have to satisfy the Hodge relations on its Betti numbers [52]. The following example first demonstrated otherwise.

The Kodaira-Thurston example (Thurston, 1976 [101]):

Take $\mathbb{R}^4$ with $dx_1 \wedge dy_1 + dx_2 \wedge dy_2$, and Γ the discrete group generated by the following symplectomorphisms:

$$\gamma_1 : (x_1,x_2,y_1,y_2) \longmapsto (x_1,x_2+1,y_1,y_2)$$
$$\gamma_2 : (x_1,x_2,y_1,y_2) \longmapsto (x_1,x_2,y_1,y_2+1)$$
$$\gamma_3 : (x_1,x_2,y_1,y_2) \longmapsto (x_1+1,x_2,y_1,y_2)$$
$$\gamma_4 : (x_1,x_2,y_1,y_2) \longmapsto (x_1,x_2+y_2,y_1+1,y_2)$$

Then $M = \mathbb{R}^4/\Gamma$ is a flat 2-torus bundle over a 2-torus. Kodaira [70] had shown that M has a complex structure. However, $\pi_1(M) = \Gamma$, hence $H^1(\mathbb{R}^4/\Gamma,\mathbb{Z}) = \Gamma/[\Gamma,\Gamma]$ has rank 3, $b^1 = 3$ is *odd*, so M is *not* Kähler [101].

- Does any symplectic manifold admit some complex structure (not necessarily compatible)?

No.

(Fernández-Gotay-Gray, 1988 [37]): There are symplectic manifolds which do *not* admit any complex structure [37]. Their examples are circle bundles over circle bundles over a 2-torus.

$$S^1 \hookrightarrow M$$
$$\downarrow$$
$$S^1 \hookrightarrow P \qquad \text{tower of circle fibrations}$$
$$\downarrow$$
$$\mathbb{T}^2$$

- Given a complex structure on M, is there always a symplectic structure (not necessarily compatible)?

No.

The **Hopf surface** $S^1 \times S^3$ is not symplectic because $H^2(S^1 \times S^3) = 0$. But it is complex since $S^1 \times S^3 \simeq \mathbb{C}^2 \backslash \{0\}/\Gamma$ where $\Gamma = \{2^n \mathrm{Id} \mid n \in \mathbb{Z}\}$ is a group of *complex* transformations, i.e., we factor $\mathbb{C}^2 \backslash \{0\}$ by the equivalence relation $(z_1, z_2) \sim (2z_1, 2z_2)$.

- Is any almost complex manifold either complex or symplectic?

No.

$\mathbb{CP}^2 \# \mathbb{CP}^2 \# \mathbb{CP}^2$ is almost complex (proved by a computation with characteristic classes), but is neither complex (since it does not fit Kodaira's classification of complex surfaces), nor symplectic (as shown by **Taubes** [97] in 1995 using Seiberg-Witten invariants).

- In 1993 **Gompf** [46] provided a construction that yields a compact symplectic 4-manifold with fundamental group equal to any given finitely-presented group. In particular, we can find simply connected examples. His construction can be adapted to produce *non*Kähler examples.

17.4 Main Kähler Manifolds

- **Compact Riemann surfaces**
  As real manifolds, these are the 2-dimensional compact orientable manifolds classified by genus. An area form is a symplectic form. Any compatible almost complex structure is always integrable for dimension reasons (see Homework 10).

- **Stein manifolds**

  Definition 17.4. A *Stein manifold* is a Kähler manifold (M, ω) which admits a global proper Kähler potential, i.e., $\omega = \frac{i}{2} \partial \bar{\partial} \rho$ for some proper function $\rho : M \to \mathbb{R}$.

  Proper means that the preimage by ρ of a compact set is compact, i.e., "$\rho(p) \to \infty$ as $p \to \infty$."

Stein manifolds can be also characterized as the properly embedded analytic sub-manifolds of $\mathbb{C}^n$.

- **Complex tori**

  Complex tori look like $M = \mathbb{C}^n/\mathbb{Z}^n$ where $\mathbb{Z}^n$ is a lattice in $\mathbb{C}^n$. The form $\omega = \sum dz_j \wedge d\bar{z}_j$ induced by the euclidean structure is Kähler.

- **Complex projective spaces**

  The standard Kähler form on $\mathbb{CP}^n$ is the Fubini-Study form (see Homework 12). (In 1995, Taubes showed that $\mathbb{CP}^2$ has a unique symplectic structure up to symplectomorphism.)

- **Products of Kähler manifolds**
- **Complex submanifolds of Kähler manifolds**

Part VII
Hamiltonian Mechanics

The equations of motion in classical mechanics arise as solutions of variational problems. For a general mechanical system of n particles in $\mathbb{R}^3$, the physical path satisfies Newton's second law. On the other hand, the physical path minimizes the mean value of kinetic minus potential energy. This quantity is called the action. For a system with constraints, the physical path is the path which minimizes the action among all paths satisfying the constraint.

The Legendre transform (Lecture 20) gives the relation between the variational (Euler-Lagrange) and the symplectic (Hamilton-Jacobi) formulations of the equations of motion.

Chapter 18
Hamiltonian Vector Fields

18.1 Hamiltonian and Symplectic Vector Fields

– What does a symplectic geometer do with a real function?...

Let (M, ω) be a symplectic manifold and let $H : M \to \mathbb{R}$ be a smooth function. Its differential dH is a 1-form. By nondegeneracy, there is a unique vector field X_H on M such that $\iota_{X_H} \omega = dH$. Integrate X_H. Supposing that M is compact, or at least that X_H is complete, let $\rho_t : M \to M$, $t \in \mathbb{R}$, be the one-parameter family of diffeomorphisms generated by X_H:

$$\begin{cases} \rho_0 = \mathrm{id}_M \\ \dfrac{d\rho_t}{dt} \circ \rho_t^{-1} = X_H \ . \end{cases}$$

Claim. Each diffeomorphism ρ_t preserves ω, i.e., $\rho_t^* \omega = \omega$, $\forall t$.

Proof. We have $\frac{d}{dt} \rho_t^* \omega = \rho_t^* \mathcal{L}_{X_H} \omega = \rho_t^* (d \underbrace{\iota_{X_H} \omega}_{dH} + \iota_{X_H} \underbrace{d\omega}_{0}) = 0.$ $\qquad \square$

Therefore, every function on (M, ω) gives a family of symplectomorphisms. Notice how the proof involved both the *nondegeneracy* and the *closedness* of ω.

Definition 18.1. A vector field X_H as above is called the **hamiltonian vector field** with **hamiltonian function** H.

Example. The height function $H(\theta, h) = h$ on the sphere $(M, \omega) = (S^2, d\theta \wedge dh)$ has

$$\iota_{X_H}(d\theta \wedge dh) = dh \quad \Longleftrightarrow \quad X_H = \frac{\partial}{\partial \theta} \ .$$

Thus, $\rho_t(\theta, h) = (\theta + t, h)$, which is rotation about the vertical axis; the height function H is preserved by this motion. $\qquad \diamondsuit$

Exercise. Let X be a vector field on an abstract manifold W. There is a unique vector field $X_\sharp$ on the cotangent bundle T^*W, whose flow is the lift of the flow of X; cf. Lecture 2. Let α be the tautological 1-form on T^*W and let $\omega = -d\alpha$ be the canonical symplectic form on T^*W. Show that $X_\sharp$ is a hamiltonian vector field with hamiltonian function $H := \iota_{X_\sharp}\alpha$. $\diamond$

Remark. If X_H is hamiltonian, then

$$\mathcal{L}_{X_H}H = \iota_{X_H}dH = \iota_{X_H}\iota_{X_H}\omega = 0 \ .$$

Therefore, hamiltonian vector fields preserve their hamiltonian functions, and each integral curve $\{\rho_t(x) \mid t \in \mathbb{R}\}$ of X_H must be contained in a level set of H:

$$H(x) = (\rho_t^*H)(x) = H(\rho_t(x)) \ , \quad \forall t \ .$$

Definition 18.2. A vector field X on M preserving ω (i.e., such that $\mathcal{L}_X\omega = 0$) is called a ***symplectic vector field***.

$$\begin{cases} X \text{ is symplectic} & \Longleftrightarrow \ \iota_X\omega \text{ is closed} \ , \\ X \text{ is hamiltonian} & \Longleftrightarrow \ \iota_X\omega \text{ is exact} \ . \end{cases}$$

Locally, on every contractible open set, every symplectic vector field is hamiltonian. If $H^1_{\text{deRham}}(M) = 0$, then globally every symplectic vector field is hamiltonian. In general, $H^1_{\text{deRham}}(M)$ measures the obstruction for symplectic vector fields to be hamiltonian.

Example. On the 2-torus $(M, \omega) = (\mathbb{T}^2, d\theta_1 \wedge d\theta_2)$, the vector fields $X_1 = \frac{\partial}{\partial \theta_1}$ and $X_2 = \frac{\partial}{\partial \theta_2}$ are symplectic but not hamiltonian. $\diamond$

To summarize, vector fields on a symplectic manifold (M, ω) which preserve ω are called **symplectic**. The following are equivalent:

- X is a symplectic vector field;
- the flow ρ_t of X preserves ω, i.e., $\rho_t^*\omega = \omega$, for all t;
- $\mathcal{L}_X\omega = 0$;
- $\iota_X\omega$ is closed.

A **hamiltonian** vector field is a vector field X for which

- $\iota_X\omega$ is exact,

i.e., $\iota_X\omega = dH$ for some $H \in C^\infty(M)$. A primitive H of $\iota_X\omega$ is then called a **hamiltonian function** of X.

18.2 Classical Mechanics

Consider euclidean space $\mathbb{R}^{2n}$ with coordinates $(q_1, \ldots, q_n, p_1, \ldots, p_n)$ and $\omega_0 = \sum dq_j \wedge dp_j$. The curve $\rho_t = (q(t), p(t))$ is an integral curve for X_H exactly if

$$
\begin{cases}
\dfrac{dq_i}{dt}(t) = \dfrac{\partial H}{\partial p_i} \\[3mm]
\dfrac{dp_i}{dt}(t) = -\dfrac{\partial H}{\partial q_i}
\end{cases}
\qquad \textbf{(Hamilton equations)}
$$

Indeed, let $X_H = \sum\limits_{i=1}^{n} \left(\dfrac{\partial H}{\partial p_i}\dfrac{\partial}{\partial q_i} - \dfrac{\partial H}{\partial q_i}\dfrac{\partial}{\partial p_i} \right)$. Then,

$$
\iota_{X_H}\omega = \sum_{j=1}^{n} \iota_{X_H}(dq_j \wedge dp_j) = \sum_{j=1}^{n} [(\iota_{X_H}dq_j) \wedge dp_j - dq_j \wedge (\iota_{X_H}dp_j)]
$$
$$
= \sum_{j=1}^{n} \left(\dfrac{\partial H}{\partial p_j}dp_j + \dfrac{\partial H}{\partial q_j}dq_j \right) = dH .
$$

Remark. The gradient vector field of H relative to the euclidean metric is

$$
\nabla H := \sum_{i=1}^{n} \left(\dfrac{\partial H}{\partial q_i}\dfrac{\partial}{\partial q_i} + \dfrac{\partial H}{\partial p_i}\dfrac{\partial}{\partial p_i} \right) .
$$

If J is the standard (almost) complex structure so that $J(\frac{\partial}{\partial q_i}) = \frac{\partial}{\partial p_i}$ and $J(\frac{\partial}{\partial p_i}) = -\frac{\partial}{\partial q_i}$, we have $JX_H = \nabla H$. $\qquad\qquad \Diamond$

The case where $n = 3$ has a simple physical illustration. Newton's second law states that a particle of mass m moving in **configuration space** $\mathbb{R}^3$ with coordinates $q = (q_1, q_2, q_3)$ under a potential $V(q)$ moves along a curve $q(t)$ such that

$$
m\dfrac{d^2q}{dt^2} = -\nabla V(q) .
$$

Introduce the **momenta** $p_i = m\frac{dq_i}{dt}$ for $i = 1, 2, 3$, and **energy** function $H(p, q) = \frac{1}{2m}|p|^2 + V(q)$. Let $\mathbb{R}^6 = T^*\mathbb{R}^3$ be the corresponding **phase space**, with coordinates $(q_1, q_2, q_3, p_1, p_2, p_3)$. Newton's second law in $\mathbb{R}^3$ is equivalent to the Hamilton equations in $\mathbb{R}^6$:

$$
\begin{cases}
\dfrac{dq_i}{dt} = \dfrac{1}{m}p_i = \dfrac{\partial H}{\partial p_i} \\[3mm]
\dfrac{dp_i}{dt} = m\dfrac{d^2q_i}{dt^2} = -\dfrac{\partial V}{\partial q_i} = -\dfrac{\partial H}{\partial q_i} .
\end{cases}
$$

The energy H is conserved by the physical motion.

18.3 Brackets

Vector fields are differential operators on functions: if X is a vector field and $f \in C^\infty(M)$, df being the corresponding 1-form, then

$$X \cdot f := df(X) = \mathcal{L}_X f \ .$$

Given two vector fields X, Y, there is a unique vector field W such that

$$\mathcal{L}_W f = \mathcal{L}_X(\mathcal{L}_Y f) - \mathcal{L}_Y(\mathcal{L}_X f) \ .$$

The vector field W is called the **Lie bracket** of the vector fields X and Y and denoted $W = [X,Y]$, since $\mathcal{L}_W = [\mathcal{L}_X, \mathcal{L}_Y]$ is the commutator.

Exercise. Check that, for any form α,

$$\iota_{[X,Y]}\alpha = \mathcal{L}_X \iota_Y \alpha - \iota_Y \mathcal{L}_X \alpha = [\mathcal{L}_X, \iota_Y]\alpha \ .$$

Since each side is an anti-derivation with respect to the wedge product, it suffices to check this formula on local generators of the exterior algebra of forms, namely functions and exact 1-forms. $\diamondsuit$

Proposition 18.3. *If X and Y are symplectic vector fields on a symplectic manifold (M, ω), then $[X,Y]$ is hamiltonian with hamiltonian function $\omega(Y,X)$.*

Proof.

$$\begin{aligned}
\iota_{[X,Y]}\omega &= \mathcal{L}_X \iota_Y \omega - \iota_Y \mathcal{L}_X \omega \\
&= d\iota_X \iota_Y \omega + \iota_X \underbrace{d\iota_Y \omega}_{0} - \iota_Y \underbrace{d\iota_X \omega}_{0} - \iota_Y \iota_X \underbrace{d\omega}_{0} \\
&= d(\omega(Y,X)) \ .
\end{aligned}$$

$\square$

A (real) **Lie algebra** is a (real) vector space $\mathfrak{g}$ together with a **Lie bracket** $[\cdot,\cdot]$, i.e., a bilinear map $[\cdot,\cdot] : \mathfrak{g} \times \mathfrak{g} \to \mathfrak{g}$ satisfying:

(a) $[x,y] = -[y,x] \ , \quad \forall x,y \in \mathfrak{g} \ ,$ (**antisymmetry**)
(b) $[x,[y,z]] + [y,[z,x]] + [z,[x,y]] = 0 \ , \quad \forall x,y,z \in \mathfrak{g} \ .$ (**Jacobi identity**)

Let

$$\begin{aligned}
\chi(M) &= \{ \text{ vector fields on } M \ \} \\
\chi^{\text{sympl}}(M) &= \{ \text{ symplectic vector fields on } M \ \} \\
\chi^{\text{ham}}(M) &= \{ \text{ hamiltonian vector fields on } M \ \} \ .
\end{aligned}$$

Corollary 18.4. *The inclusions $(\chi^{\text{ham}}(M), [\cdot,\cdot]) \subseteq (\chi^{\text{sympl}}(M), [\cdot,\cdot]) \subseteq (\chi(M), [\cdot,\cdot])$ are inclusions of Lie algebras.*

Definition 18.5. *The* ***Poisson bracket*** *of two functions $f, g \in C^\infty(M; \mathbb{R})$ is*

$$\{f,g\} := \omega(X_f, X_g) \ .$$

We have $X_{\{f,g\}} = -[X_f, X_g]$ because $X_{\omega(X_f,X_g)} = [X_g, X_f]$.

Theorem 18.6. *The bracket* $\{\cdot,\cdot\}$ *satisfies the Jacobi identity, i.e.,*

$$\{f,\{g,h\}\} + \{g,\{h,f\}\} + \{h,\{f,g\}\} = 0 .$$

Proof. Exercise. □

Definition 18.7. A *Poisson algebra* $(\mathcal{P}, \{\cdot,\cdot\})$ is a commutative associative algebra $\mathcal{P}$ with a Lie bracket $\{\cdot,\cdot\}$ satisfying the *Leibniz rule*:

$$\{f,gh\} = \{f,g\}h + g\{f,h\} .$$

Exercise. Check that the Poisson bracket $\{\cdot,\cdot\}$ defined above satisfies the Leibniz rule. ◇

We conclude that, if (M,ω) is a symplectic manifold, then $(C^\infty(M), \{\cdot,\cdot\})$ is a Poisson algebra. Furthermore, we have a Lie algebra anti-homomorphism

$$\begin{aligned} C^\infty(M) &\longrightarrow \chi(M) \\ H &\longmapsto X_H \\ \{\cdot,\cdot\} &\rightsquigarrow -[\cdot,\cdot] . \end{aligned}$$

18.4 Integrable Systems

Definition 18.8. A *hamiltonian system* is a triple (M,ω,H), where (M,ω) is a symplectic manifold and $H \in C^\infty(M;\mathbb{R})$ is a function, called the *hamiltonian function*.

Theorem 18.9. *We have* $\{f,H\} = 0$ *if and only if* f *is constant along integral curves of* X_H.

Proof. Let ρ_t be the flow of X_H. Then

$$\begin{aligned} \tfrac{d}{dt}(f \circ \rho_t) &= \rho_t^* \mathcal{L}_{X_H} f = \rho_t^* \iota_{X_H} df = \rho_t^* \iota_{X_H} \iota_{X_f} \omega \\ &= \rho_t^* \omega(X_f, X_H) = \rho_t^* \{f,H\} . \end{aligned}$$

□

A function f as in Theorem 18.9 is called an **integral of motion** (or a **first integral** or a **constant of motion**). In general, hamiltonian systems do not admit integrals of motion which are *independent* of the hamiltonian function. Functions $f_1,\ldots,f_n$ on M are said to be **independent** if their differentials $(df_1)_p,\ldots,(df_n)_p$ are linearly independent at all points p in some open dense subset of M. Loosely speaking, a hamiltonian system is *(completely) integrable* if it has as many commuting integrals of motion as possible. Commutativity is with respect to the Poisson bracket. Notice that, if $f_1, \ldots, f_n$ are commuting integrals of motion for a hamiltonian system (M, ω, H), then, at each $p \in M$, their hamiltonian vector fields generate an isotropic subspace of $T_p M$:

$$\omega(X_{f_i}, X_{f_j}) = \{f_i, f_j\} = 0 .$$

If $f_1, \ldots, f_n$ are independent, then, by symplectic linear algebra, n can be at most half the dimension of M.

Definition 18.10. A hamiltonian system (M, ω, H) is *(completely) integrable* if it possesses $n = \frac{1}{2} \dim M$ independent integrals of motion, $f_1 = H, f_2, \ldots, f_n$, which are pairwise in involution with respect to the Poisson bracket, i.e., $\{f_i, f_j\} = 0$, for all i, j.

Example. The simple pendulum (Homework 13) and the harmonic oscillator are trivially integrable systems – any 2-dimensional hamiltonian system (where the set of non-fixed points is dense) is integrable. $\diamond$

Example. A hamiltonian system (M, ω, H) where M is 4-dimensional is integrable if there is an integral of motion independent of H (the commutativity condition is automatically satisfied). Homework 18 shows that the spherical pendulum is integrable. $\diamond$

For sophisticated examples of integrable systems, see [10, 62].

Let (M, ω, H) be an integrable system of dimension $2n$ with integrals of motion $f_1 = H, f_2, \ldots, f_n$. Let $c \in \mathbb{R}^n$ be a regular value of $f := (f_1, \ldots, f_n)$. The corresponding level set, $f^{-1}(c)$, is a lagrangian submanifold, because it is n-dimensional and its tangent bundle is isotropic.

Lemma 18.11. *If the hamiltonian vector fields $X_{f_1}, \ldots, X_{f_n}$ are complete on the level $f^{-1}(c)$, then the connected components of $f^{-1}(c)$ are homogeneous spaces for $\mathbb{R}^n$, i.e., are of the form $\mathbb{R}^{n-k} \times \mathbb{T}^k$ for some k, $0 \leq k \leq n$, where $\mathbb{T}^k$ is a k-dimensional torus.*

Proof. Exercise (just follow the flows to obtain coordinates). $\square$

Any compact component of $f^{-1}(c)$ must hence be a torus. These components, when they exist, are called **Liouville tori**. (The easiest way to ensure that compact components exist is to have one of the f_i's proper.)

Theorem 18.12. (Arnold-Liouville [3]) *Let (M, ω, H) be an integrable system of dimension $2n$ with integrals of motion $f_1 = H, f_2, \ldots, f_n$. Let $c \in \mathbb{R}^n$ be a regular value of $f := (f_1, \ldots, f_n)$. The corresponding level $f^{-1}(c)$ is a lagrangian submanifold of M.*

(a) *If the flows of $X_{f_1}, \ldots, X_{f_n}$ starting at a point $p \in f^{-1}(c)$ are complete, then the connected component of $f^{-1}(c)$ containing p is a homogeneous space for $\mathbb{R}^n$. With respect to this affine structure, that component has coordinates $\varphi_1, \ldots, \varphi_n$, known as **angle coordinates**, in which the flows of the vector fields $X_{f_1}, \ldots, X_{f_n}$ are linear.*

(b) *There are coordinates $\psi_1, \ldots, \psi_n$, known as **action coordinates**, complementary to the angle coordinates such that the ψ_i's are integrals of motion and $\varphi_1, \ldots, \varphi_n, \psi_1, \ldots, \psi_n$ form a Darboux chart.*

Therefore, the dynamics of an integrable system is extremely simple and the system has an explicit solution in action-angle coordinates. The proof of part (a) – the easy part – of the Arnold-Liouville theorem is sketched above. For the proof of part (b), see [3, 28].

Geometrically, regular levels being lagrangian submanifolds implies that, in a neighborhood of a regular value, the map $f : M \to \mathbb{R}^n$ collecting the given integrals of motion is a **lagrangian fibration**, i.e., it is locally trivial and its fibers are lagrangian submanifolds. Part (a) of the Arnold-Liouville theorem states that there are coordinates along the fibers, the angle coordinates φ_i,[1] in which the flows of $X_{f_1}, \ldots, X_{f_n}$ are linear. Part (b) of the theorem guarantees the existence of coordinates on $\mathbb{R}^n$, the action coordinates ψ_i, which (Poisson) commute among themselves and satisfy $\{\varphi_i, \psi_j\} = \delta_{ij}$ with respect to the angle coordinates. Notice that, in general, the action coordinates are not the given integrals of motion because $\varphi_1, \ldots, \varphi_n, f_1, \ldots, f_n$ do not form a Darboux chart.

---

[1] The name "angle coordinates" is used even if the fibers are not tori.

Homework 13: Simple Pendulum

This problem is adapted from [53].

   The **simple pendulum** is a mechanical system consisting of a massless rigid rod of length l, fixed at one end, whereas the other end has a plumb bob of mass m, which may oscillate in the vertical plane. Assume that the force of gravity is constant pointing vertically downwards, and that this is the only external force acting on this system.

(a) Let θ be the oriented angle between the rod (regarded as a line segment) and the vertical direction. Let ξ be the coordinate along the fibers of T^*S^1 induced by the standard angle coordinate on S^1. Show that the function $H : T^*S^1 \to \mathbb{R}$ given by

$$H(\theta, \xi) = \underbrace{\frac{\xi^2}{2ml^2}}_{K} + \underbrace{ml(1 - \cos\theta)}_{V} ,$$

   is an appropriate hamiltonian function to describe the simple pendulum. More precisely, check that gravity corresponds to the potential energy $V(\theta) = ml(1 - \cos\theta)$ (we omit universal constants), and that the kinetic energy is given by $K(\theta, \xi) = \frac{1}{2ml^2}\xi^2$.

(b) For simplicity assume that $m = l = 1$.
   Plot the level curves of H in the (θ, ξ) plane.
   Show that there exists a number c such that for $0 < h < c$ the level curve $H = h$ is a disjoint union of closed curves. Show that the projection of each of these curves onto the θ-axis is an interval of length less than π.
   Show that neither of these assertions is true if $h > c$.
   What types of motion are described by these two types of curves?
   What about the case $H = c$?

(c) Compute the critical points of the function H. Show that, modulo 2π in θ, there are exactly two critical points: a critical point s where H vanishes, and a critical point u where H equals c. These points are called the **stable** and **unstable** points of H, respectively. Justify this terminology, i.e., show that a trajectory of the hamiltonian vector field of H whose initial point is close to s stays close to s forever, and show that this is not the case for u. What is happening physically?

Chapter 19
Variational Principles

19.1 Equations of Motion

The equations of motion in classical mechanics arise as solutions of variational problems:

> A general mechanical system possesses both kinetic and potential energy. The quantity that is minimized is the mean value of kinetic minus potential energy.

Example. Suppose that a point-particle of mass m moves in $\mathbb{R}^3$ under a force field F; let $x(t)$, $a \leq t \leq b$, be its path of motion in $\mathbb{R}^3$. Newton's second law states that

$$m\frac{d^2x}{dt^2}(t) = F(x(t)) \, .$$

Define the **work** of a path $\gamma : [a,b] \longrightarrow \mathbb{R}^3$, with $\gamma(a) = p$ and $\gamma(b) = q$, to be

$$W_\gamma = \int_a^b F(\gamma(t)) \cdot \frac{d\gamma}{dt}(t)dt \, .$$

Suppose that F is **conservative**, i.e., W_γ depends only on p and q. Then we can define the **potential energy** $V : \mathbb{R}^3 \longrightarrow \mathbb{R}$ of the system as

$$V(q) := W_\gamma$$

where γ is a path joining a fixed base point $p_0 \in \mathbb{R}^3$ (the "origin") to q. Newton's second law can now be written

$$m\frac{d^2x}{dt^2}(t) = -\frac{\partial V}{\partial x}(x(t)) \, .$$

In the previous lecture we saw that

$$\begin{array}{ccc}
\text{Newton's second law} & \Longleftrightarrow & \text{Hamilton equations} \\
\text{in } \mathbb{R}^3 = \{(q_1, q_2, q_3)\} & & \text{in } T^*\mathbb{R}^3 = \{(q_1, q_2, q_3, p_1, p_2, p_3)\}
\end{array}$$

where $p_i = m\frac{dq_i}{dt}$ and the hamiltonian is $H(p,q) = \frac{1}{2m}|p|^2 + V(q)$. Hence, solving Newton's second law in **configuration space** $\mathbb{R}^3$ is equivalent to solving in **phase space** $T^*\mathbb{R}^3$ for the integral curve of the hamiltonian vector field with hamiltonian function H. $\diamond$

Example. The motion of earth about the sun, both regarded as point-masses and assuming that the sun to be stationary at the origin, obeys the **inverse square law**

$$m\frac{d^2x}{dt^2} = -\frac{\partial V}{\partial x} \, ,$$

where $x(t)$ is the position of earth at time t, and $V(x) = \frac{\text{const.}}{|x|}$ is the **gravitational potential**. $\diamond$

19.2 Principle of Least Action

When we need to deal with systems with constraints, such as the simple pendulum, or two point masses attached by a rigid rod, or a rigid body, the language of variational principles becomes more appropriate than the explicit analogues of Newton's second laws. Variational principles are due mostly to D'Alembert, Maupertius, Euler and Lagrange.

Example. (The n-particle system.) Suppose that we have n point-particles of masses $m_1, \ldots, m_n$ moving in 3-space. At any time t, the configuration of this system is described by a vector in configuration space $\mathbb{R}^{3n}$

$$x = (x_1, \ldots, x_n) \in \mathbb{R}^{3n}$$

with $x_i \in \mathbb{R}^3$ describing the position of the ith particle. If $V \in C^\infty(\mathbb{R}^{3n})$ is the potential energy, then a path of motion $x(t)$, $a \leq t \leq b$, satisfies

$$m_i\frac{d^2x_i}{dt^2}(t) = -\frac{\partial V}{\partial x_i}(x_1(t), \ldots, x_n(t)).$$

Consider this path in configuration space as a map $\gamma_0 : [a,b] \to \mathbb{R}^{3n}$ with $\gamma_0(a) = p$ and $\gamma_0(b) = q$, and let

$$\mathcal{P} = \{\gamma : [a,b] \longrightarrow \mathbb{R}^{3n} \mid \gamma(a) = p \text{ and } \gamma(b) = q\}$$

be the set of all paths going from p to q over time $t \in [a,b]$. $\diamond$

Definition 19.1. The *action* of a path $\gamma \in \mathcal{P}$ is

$$\mathcal{A}_\gamma := \int_a^b \left(\sum_{i=1}^n \frac{m_i}{2}\left|\frac{d\gamma_i}{dt}(t)\right|^2 - V(\gamma(t))\right) dt \, .$$

Principle of least action.

The physical path γ_0 is the path for which $\mathcal{A}_\gamma$ is minimal.

Newton's second law for a constrained system.

Suppose that the n point-masses are restricted to move on a submanifold M of $\mathbb{R}^{3n}$ called the **constraint set**. We can now single out the actual physical path $\gamma_0 :$ $[a,b] \to M$, with $\gamma_0(a) = p$ and $\gamma_0(b) = q$, as being "the" path which minimizes $\mathcal{A}_\gamma$ among all those hypothetical paths $\gamma : [a,b] \to \mathbb{R}^{3n}$ with $\gamma(a) = p$, $\gamma(b) = q$ and satisfying the rigid constraints $\gamma(t) \in M$ for all t.

19.3 Variational Problems

Let M be an n-dimensional manifold. Its tangent bundle TM is a $2n$-dimensional manifold. Let $F : TM \to \mathbb{R}$ be a smooth function.

If $\gamma : [a,b] \to M$ is a smooth curve on M, define the **lift of γ to TM** to be the smooth curve on TM given by

$$\tilde{\gamma} : [a,b] \longrightarrow TM$$
$$t \longmapsto \left(\gamma(t), \frac{d\gamma}{dt}(t) \right) .$$

The **action** of γ is

$$\mathcal{A}_\gamma := \int_a^b (\tilde{\gamma}^* F)(t) dt = \int_a^b F\left(\gamma(t), \frac{d\gamma}{dt}(t) \right) dt .$$

For fixed $p, q \in M$, let

$$\mathcal{P}(a,b,p,q) := \{\gamma : [a,b] \longrightarrow M \mid \gamma(a) = p, \ \gamma(b) = q\} .$$

Problem.

Find, among all $\gamma \in \mathcal{P}(a,b,p,q)$, the curve γ_0 which "minimizes" $\mathcal{A}_\gamma$.

First observe that minimizing curves are always locally minimizing:

Lemma 19.2. *Suppose that $\gamma_0 : [a,b] \to M$ is minimizing. Let $[a_1,b_1]$ be a subinterval of $[a,b]$ and let $p_1 = \gamma_0(a_1)$, $q_1 = \gamma_0(b_1)$. Then $\gamma_0|_{[a_1,b_1]}$ is minimizing among the curves in $\mathcal{P}(a_1,b_1,p_1,q_1)$.*

Proof. Exercise:

Argue by contradiction. Suppose that there were $\gamma_1 \in \mathcal{P}(a_1,b_1,p_1,q_1)$ for which $\mathcal{A}_{\gamma_1} < \mathcal{A}_{\gamma_0|_{[a_1,b_1]}}$. Consider a broken path obtained from γ_0 by replacing the segment $\gamma_0|_{[a_1,b_1]}$ by γ_1. Construct a smooth curve $\gamma_2 \in \mathcal{P}(a,b,p,q)$ for which $\mathcal{A}_{\gamma_2} < \mathcal{A}_{\gamma_0}$ by rounding off the corners of the broken path. $\square$

We will now assume that p, q and γ_0 lie in a coordinate neighborhood $(\mathcal{U}, x_1, \ldots, x_n)$. On $T\mathcal{U}$ we have coordinates $(x_1, \ldots, x_n, v_1, \ldots, v_n)$ associated with a trivialization of $T\mathcal{U}$ by $\frac{\partial}{\partial x_1}, \ldots, \frac{\partial}{\partial x_n}$. Using this trivialization, the curve

$$\gamma : [a, b] \longrightarrow \mathcal{U}, \qquad \gamma(t) = (\gamma_1(t), \ldots, \gamma_n(t))$$

lifts to

$$\tilde{\gamma} : [a, b] \longrightarrow T\mathcal{U}, \qquad \tilde{\gamma}(t) = \left(\gamma_1(t), \ldots, \gamma_n(t), \frac{d\gamma_1}{dt}(t), \ldots, \frac{d\gamma_n}{dt}(t) \right).$$

Necessary condition for $\gamma_0 \in \mathcal{P}(a, b, p, q)$ to minimize the action.

Let $c_1, \ldots, c_n \in C^\infty([a, b])$ be such that $c_i(a) = c_i(b) = 0$. Let $\gamma_\varepsilon : [a, b] \longrightarrow \mathcal{U}$ be the curve

$$\gamma_\varepsilon(t) = (\gamma_1(t) + \varepsilon c_1(t), \ldots, \gamma_n(t) + \varepsilon c_n(t)).$$

For ε small, γ_ε is well-defined and in $\mathcal{P}(a, b, p, q)$.

Let $\mathcal{A}_\varepsilon = \mathcal{A}_{\gamma_\varepsilon} = \int_a^b F\left(\gamma_\varepsilon(t), \frac{d\gamma_\varepsilon}{dt}(t) \right) dt$. If γ_0 minimizes $\mathcal{A}$, then

$$\frac{d\mathcal{A}_\varepsilon}{d\varepsilon}(0) = 0.$$

$$\frac{d\mathcal{A}_\varepsilon}{d\varepsilon}(0) = \int_a^b \sum_i \left[\frac{\partial F}{\partial x_i}\left(\gamma_0(t), \frac{d\gamma_0}{dt}(t) \right) c_i(t) + \frac{\partial F}{\partial v_i}\left(\gamma_0(t), \frac{d\gamma_0}{dt}(t) \right) \frac{dc_i}{dt}(t) \right] dt$$

$$= \int_a^b \sum_i \left[\frac{\partial F}{\partial x_i}(\ldots) - \frac{d}{dt}\frac{\partial F}{\partial v_i}(\ldots) \right] c_i(t) dt = 0$$

where the first equality follows from the Leibniz rule and the second equality follows from integration by parts. Since this is true for all c_i's satisfying the boundary conditions $c_i(a) = c_i(b) = 0$, we conclude that

$$\frac{\partial F}{\partial x_i}\left(\gamma_0(t), \frac{d\gamma_0}{dt}(t) \right) = \frac{d}{dt}\frac{\partial F}{\partial v_i}\left(\gamma_0(t), \frac{d\gamma_0}{dt}(t) \right). \qquad \textbf{E-L}$$

These are the **Euler-Lagrange equations**.

19.4 Solving the Euler-Lagrange Equations

Case 1: Suppose that $F(x, v)$ does not depend on v.

The Euler-Lagrange equations become

$$\frac{\partial F}{\partial x_i}\left(\gamma_0(t), \frac{d\gamma_0}{dt}(t) \right) = 0 \iff \text{the curve } \gamma_0 \text{ sits on the critical set of } F.$$

For generic F, the critical points are isolated, hence $\gamma_0(t)$ must be a constant curve.

Case 2: Suppose that $F(x,v)$ depends affinely on v:

$$F(x,v) = F_0(x) + \sum_{j=1}^{n} F_j(x)v_j .$$

LHS of **E-L**: $\qquad \dfrac{\partial F_0}{\partial x_i}(\gamma(t)) + \sum_{j=1}^{n} \dfrac{\partial F_j}{\partial x_i}(\gamma(t))\dfrac{d\gamma_j}{dt}(t)$

RHS of **E-L**: $\qquad \dfrac{d}{dt}F_i(\gamma(t)) = \sum_{j=1}^{n} \dfrac{\partial F_i}{\partial x_j}(\gamma(t))\dfrac{d\gamma_j}{dt}(t)$

The Euler-Lagrange equations become

$$\frac{\partial F_0}{\partial x_i}(\gamma(t)) = \sum_{j=1}^{n} \underbrace{\left(\frac{\partial F_i}{\partial x_j} - \frac{\partial F_j}{\partial x_i}\right)}_{n\times n \text{ matrix}}(\gamma(t))\frac{d\gamma_j}{dt}(t) .$$

If the $n \times n$ matrix $\left(\frac{\partial F_i}{\partial x_j} - \frac{\partial F_j}{\partial x_i}\right)$ has an inverse $G_{ij}(x)$, then

$$\frac{d\gamma_j}{dt}(t) = \sum_{i=1}^{n} G_{ji}(\gamma(t))\frac{\partial F_0}{\partial x_i}(\gamma(t))$$

is a system of first order ordinary differential equations. Locally it has a unique solution through each point p. If q is not on this curve, there is no solution at all to the Euler-Lagrange equations belonging to $\mathcal{P}(a,b,p,q)$.

Therefore, we need non-linear dependence of F on the v variables in order to have appropriate solutions. From now on, assume that the

Legendre condition: $\qquad \det\left(\dfrac{\partial^2 F}{\partial v_i \partial v_j}\right) \neq 0 .$

Letting $G_{ij}(x,v) = \left(\dfrac{\partial^2 F}{\partial v_i \partial v_j}(x,v)\right)^{-1}$, the Euler-Lagrange equations become

$$\frac{d^2\gamma_j}{dt^2} = \sum_{i} G_{ji}\frac{\partial F}{\partial x_i}\left(\gamma,\frac{d\gamma}{dt}\right) - \sum_{i,k} G_{ji}\frac{\partial^2 F}{\partial v_i \partial x_k}\left(\gamma,\frac{d\gamma}{dt}\right)\frac{d\gamma_k}{dt} .$$

This second order ordinary differential equation has a unique solution given initial conditions

$$\gamma(a) = p \qquad \text{and} \qquad \frac{d\gamma}{dt}(a) = v .$$

19.5 Minimizing Properties

Is the above solution locally minimizing?

Assume that $\left(\frac{\partial^2 F}{\partial v_i \partial v_j}(x,v)\right) \gg 0$, $\forall (x,v)$, i.e., with the x variable frozen, the function $v \mapsto F(x,v)$ is **strictly convex**.

Suppose that $\gamma_0 \in \mathcal{P}(a,b,p,q)$ satisfies **E-L**. Does γ_0 minimize $\mathcal{A}_\gamma$? Locally, yes, according to the following theorem. (Globally it is only critical.)

Theorem 19.3. *For every sufficiently small subinterval $[a_1,b_1]$ of $[a,b]$, $\gamma_0|_{[a_1,b_1]}$ is locally minimizing in $\mathcal{P}(a_1,b_1,p_1,q_1)$ where $p_1 = \gamma_0(a_1)$, $q_1 = \gamma_0(b_1)$.*

Proof. As an exercise in Fourier series, show the **Wirtinger inequality**: for $f \in C^1([a,b])$ with $f(a) = f(b) = 0$, we have

$$\int_a^b \left|\frac{df}{dt}\right|^2 dt \geq \frac{\pi^2}{(b-a)^2} \int_a^b |f|^2 dt \ .$$

Suppose that $\gamma_0 : [a,b] \to \mathcal{U}$ satisfies **E-L**. Take $c_i \in C^\infty([a,b])$, $c_i(a) = c_i(b) = 0$. Let $c = (c_1,\dots,c_n)$. Let $\gamma_\varepsilon = \gamma_0 + \varepsilon c \in \mathcal{P}(a,b,p,q)$, and let $\mathcal{A}_\varepsilon = \mathcal{A}_{\gamma_\varepsilon}$.

E-L $\iff \frac{d\mathcal{A}_\varepsilon}{d\varepsilon}(0) = 0$.

$$\frac{d^2 \mathcal{A}_\varepsilon}{d\varepsilon^2}(0) = \int_a^b \sum_{i,j} \frac{\partial^2 F}{\partial x_i \partial x_j}\left(\gamma_0, \frac{d\gamma_0}{dt}\right) c_i c_j \, dt \qquad \text{(I)}$$

$$+ 2\int_a^b \sum_{i,j} \frac{\partial^2 F}{\partial x_i \partial v_j}\left(\gamma_0, \frac{d\gamma_0}{dt}\right) c_i \frac{dc_j}{dt} \, dt \qquad \text{(II)}$$

$$+ \int_a^b \sum_{i,j} \frac{\partial^2 F}{\partial v_i \partial v_j}\left(\gamma_0, \frac{d\gamma_0}{dt}\right) \frac{dc_i}{dt} \frac{dc_j}{dt} \, dt \qquad \text{(III)} \ .$$

Since $\left(\frac{\partial^2 F}{\partial v_i \partial v_j}(x,v)\right) \gg 0$ at all x,v,

$$\text{III} \geq K_{\text{III}} \left|\frac{dc}{dt}\right|^2_{L^2[a,b]}$$

$$|\text{I}| \leq K_{\text{I}} |c|^2_{L^2[a,b]}$$

$$|\text{II}| \leq K_{\text{II}} |c|_{L^2[a,b]} \left|\frac{dc}{dt}\right|_{L^2[a,b]}$$

where $K_{\text{I}}, K_{\text{II}}, K_{\text{III}} > 0$. By the Wirtinger inequality, if $b - a$ is very small, then $\text{III} > |\text{I}| + |\text{II}|$ when $c \neq 0$. Hence, γ_0 is a local minimum. $\square$

Homework 14: Minimizing Geodesics

This set of problems is adapted from [53].

Let (M, g) be a riemannian manifold. From the riemannian metric, we get a function $F : TM \to \mathbb{R}$, whose restriction to each tangent space T_pM is the quadratic form defined by the metric.

Let p and q be points on M, and let $\gamma : [a, b] \to M$ be a smooth curve joining p to q. Let $\tilde{\gamma} : [a, b] \to TM$, $\tilde{\gamma}(t) = (\gamma(t), \frac{d\gamma}{dt}(t))$ be the lift of γ to TM. The **action** of γ is

$$\mathcal{A}(\gamma) = \int_a^b (\tilde{\gamma}^* F) \, dt = \int_a^b \left| \frac{d\gamma}{dt} \right|^2 dt \; .$$

1. Let $\gamma : [a, b] \to M$ be a smooth curve joining p to q. Show that the arc-length of γ is independent of the parametrization of γ, i.e., show that if we reparametrize γ by $\tau : [a', b'] \to [a, b]$, the new curve $\gamma' = \gamma \circ \tau : [a', b'] \to M$ has the same arc-length.
2. Show that, given any curve $\gamma : [a, b] \to M$ (with $\frac{d\gamma}{dt}$ never vanishing), there is a reparametrization $\tau : [a, b] \to [a, b]$ such that $\gamma \circ \tau : [a, b] \to M$ is of constant velocity, that is, $|\frac{d\gamma}{dt}|$ is independent of t.
3. Let $\tau : [a, b] \to [a, b]$ be a smooth monotone map taking the endpoints of $[a, b]$ to the endpoints of $[a, b]$. Prove that

$$\int_a^b \left(\frac{d\tau}{dt} \right)^2 dt \geq b - a \; ,$$

   with equality holding if and only if $\frac{d\tau}{dt} = 1$.
4. Let $\gamma : [a, b] \to M$ be a smooth curve joining p to q. Suppose that, as s goes from a to b, its image $\gamma(s)$ moves at constant velocity, i.e., suppose that $|\frac{d\gamma}{ds}|$ is constant as a function of s. Let $\gamma' = \gamma \circ \tau : [a, b] \to M$ be a reparametrization of γ. Show that $\mathcal{A}(\gamma') \geq \mathcal{A}(\gamma)$, with equality holding if and only if $\tau(t) \equiv t$.
5. Let $\gamma_0 : [a, b] \to M$ be a curve joining p to q. Suppose that γ_0 is **action-minimizing**, i.e., suppose that

$$\mathcal{A}(\gamma_0) \leq \mathcal{A}(\gamma)$$

   for any other curve $\gamma : [a, b] \to M$ joining p to q. Prove that γ_0 is also **arc-length-minimizing**, i.e., show that γ_0 is the shortest geodesic joining p to q.
6. Show that, among all curves joining p to q, γ_0 minimizes the action if and only if γ_0 is of constant velocity and γ_0 minimizes arc-length.
7. On a coordinate chart $(\mathcal{U}, x^1, \ldots, x^n)$ on M, we have

$$F(x, v) = \sum g_{ij}(x) v^i v^j \; .$$

Show that the Euler-Lagrange equations associated to the action reduce to the **Christoffel equations** for a geodesic

$$\frac{d^2\gamma^k}{dt^2} + \sum (\Gamma^k_{ij} \circ \gamma)\frac{d\gamma^i}{dt}\frac{d\gamma^j}{dt} = 0,$$

where the Γ^k_{ij}'s (called the **Christoffel symbols**) are defined in terms of the coefficients of the riemannian metric by

$$\Gamma^k_{ij} = \frac{1}{2}\sum_\ell g^{\ell k}\left(\frac{\partial g_{\ell i}}{\partial x_j} + \frac{\partial g_{\ell j}}{\partial x_i} - \frac{\partial g_{ij}}{\partial x_\ell}\right),$$

(g^{ij}) being the matrix inverse to (g_{ij}).

8. Let p and q be two non-antipodal points on S^n. Show that the geodesic joining p to q is an arc of a great circle, the great circle in question being the intersection of S^n with the two-dimensional subspace of $\mathbb{R}^{n+1}$ spanned by p and q.

> **Hint:** No calculations are needed: Show that an isometry of a riemannian manifold has to carry geodesics into geodesics, and show that there is an isometry of $\mathbb{R}^{n+1}$ whose fixed point set is the plane spanned by p and q, and show that this isometry induces on S^n an isometry whose fixed point set is the great circle containing p and q.

Chapter 20
Legendre Transform

20.1 Strict Convexity

Let V be an n-dimensional vector space, with $e_1, \ldots, e_n$ a basis of V and $v_1, \ldots, v_n$ the associated coordinates. Let $F : V \to \mathbb{R}$, $F = F(v_1, \ldots, v_n)$, be a smooth function. Let $p \in V, u = \sum_{i=1}^{n} u_i e_i \in V$. The **hessian** of F is the quadratic function on V defined by

$$(d^2 F)_p(u) := \sum_{i,j} \frac{\partial^2 F}{\partial v_i \partial v_j}(p) u_i u_j .$$

Exercise. Show that $(d^2 F)_p(u) = \frac{d^2}{dt^2} F(p + tu)|_{t=0}$. ◇

Definition 20.1. The function F is ***strictly convex*** if $(d^2 F)_p \gg 0, \forall p \in V$.

Proposition 20.2. *For a strictly convex function F on V, the following are equivalent:*

(a) F has a critical point, i.e., a point where $dF_p = 0$;
(b) F has a local minimum at some point;
(c) F has a unique critical point (global minimum); and
(d) F is proper, that is, $F(p) \to +\infty$ as $p \to \infty$ in V.

Proof. Homework 15. □

Definition 20.3. A strictly convex function F is ***stable*** when it satisfies conditions (a)-(d) in Proposition 20.2.

Example. The function $e^x + ax$ is strictly convex for any $a \in \mathbb{R}$, but it is stable only for $a < 0$. The function $x^2 + ax$ is strictly convex and stable for any $a \in \mathbb{R}$. ◇

20.2 Legendre Transform

Let F be any strictly convex function on V. Given $\ell \in V^*$, let

$$F_\ell : V \longrightarrow \mathbb{R} , \qquad F_\ell(v) = F(v) - \ell(v) .$$

Since $(d^2 F)_p = (d^2 F_\ell)_p$,

$$F \text{ is strictly convex} \quad \Longleftrightarrow \quad F_\ell \text{ is strictly convex.}$$

Definition 20.4. The *stability set* of a strictly convex function F is

$$S_F = \{\ell \in V^* \mid F_\ell \text{ is stable}\} .$$

Proposition 20.5. *The set S_F is an open and convex subset of V^*.*

Proof. Homework 15.   □

Homework 15 also describes a sufficient condition for $S_F = V^*$.

Definition 20.6. The *Legendre transform* associated to $F \in C^\infty(V; \mathbb{R})$ is the map

$$\begin{aligned} L_F : V &\longrightarrow V^* \\ p &\longmapsto dF_p \in T_p^* V \simeq V^* . \end{aligned}$$

Proposition 20.7. *Suppose that F is strictly convex. Then*

$$L_F : V \xrightarrow{\simeq} S_F ,$$

i.e., L_F is a diffeomorphism onto S_F.

The inverse map $L_F^{-1} : S_F \to V$ is described as follows: for $\ell \in S_F$, the value $L_F^{-1}(\ell)$ is the unique minimum point $p_\ell \in V$ of $F_\ell = F - \ell$.

Exercise. Check that p is the minimum of $F(v) - dF_p(v)$.   ◇

Definition 20.8. The *dual function* F^* to F is

$$F^* : S_F \longrightarrow \mathbb{R} , \quad F^*(\ell) = - \min_{p \in V} F_\ell(p) .$$

Theorem 20.9. *We have that* $\quad L_F^{-1} = L_{F^*} .$

Proof. Homework 15.   □

20.3 Application to Variational Problems

Let M be a manifold and $F : TM \to \mathbb{R}$ a function on TM.

Problem. Minimize $\mathcal{A}_\gamma = \int \tilde{\gamma}^* F$.

At $p \in M$, let

$$F_p := F|_{T_p M} : T_p M \longrightarrow \mathbb{R} .$$

Assume that F_p is strictly convex for all $p \in M$. To simplify notation, assume also that $S_{F_p} = T_p^* M$. The Legendre transform on each tangent space

$$L_{F_p} : T_p M \xrightarrow{\simeq} T_p^* M$$

is essentially given by the first derivatives of F in the v directions. The dual function to F_p is $F_p^* : T_p^* M \longrightarrow \mathbb{R}$. Collect these fiberwise maps into

$$\mathcal{L} : TM \longrightarrow T^* M , \qquad \mathcal{L}|_{T_p M} = L_{F_p} , \qquad \text{and}$$

$$H : T^* M \longrightarrow \mathbb{R} , \qquad H|_{T_p^* M} = F_p^* .$$

Exercise. The maps H and $\mathcal{L}$ are smooth, and $\mathcal{L}$ is a diffeomorphism. $\diamond$

Let

$$\gamma : [a,b] \longrightarrow M \qquad \text{be a curve,} \quad \text{and}$$
$$\tilde{\gamma} : [a,b] \longrightarrow TM \qquad \text{its lift.}$$

Theorem 20.10. *The curve γ satisfies the Euler-Lagrange equations on every coordinate chart if and only if $\mathcal{L} \circ \tilde{\gamma} : [a,b] \to T^* M$ is an integral curve of the hamiltonian vector field X_H.*

Proof. Let

$$(\mathcal{U}, x_1, \ldots, x_n) \qquad \text{coordinate neighborhood in } M ,$$
$$(T\mathcal{U}, x_1, \ldots, x_n, v_1, \ldots, v_n) \qquad \text{coordinates in } TM ,$$
$$(T^* \mathcal{U}, x_1, \ldots, x_n, \xi_1, \ldots, \xi_n) \qquad \text{coordinates in } T^* M .$$

On $T\mathcal{U}$ we have $F = F(x, v)$.
On $T^* \mathcal{U}$ we have $H = H(u, \xi)$.

$$\mathcal{L} : \quad T\mathcal{U} \longrightarrow T^* \mathcal{U}$$
$$(x, v) \longmapsto (x, \xi) \qquad \text{where} \qquad \xi - L_{F_x}(v) = \frac{\partial F}{\partial v}(x, v) .$$

(This is the definition of **momentum** ξ.)

$$H(x, \xi) = F_x^*(\xi) = \xi \cdot v - F(x, v) \qquad \text{where} \qquad \mathcal{L}(x, v) = (x, \xi) .$$

Integral curves $(x(t), \xi(t))$ of X_H satisfy the Hamilton equations:

$$\mathbf{H} \qquad \begin{cases} \dfrac{dx}{dt} = \dfrac{\partial H}{\partial \xi}(x, \xi) \\[3mm] \dfrac{d\xi}{dt} = -\dfrac{\partial H}{\partial x}(x, \xi) , \end{cases}$$

whereas the physical path $x(t)$ satisfies the Euler-Lagrange equations:

$$\mathbf{E\text{-}L} \qquad \frac{\partial F}{\partial x}\left(x, \frac{dx}{dt}\right) = \frac{d}{dt}\frac{\partial F}{\partial v}\left(x, \frac{dx}{dt}\right) .$$

Let $(x(t), \xi(t)) = \mathcal{L}\left(x(t), \frac{dx}{dt}(t)\right)$. We want to prove:

$$t \mapsto (x(t), \xi(t)) \text{ satisfies } \mathbf{H} \quad \Longleftrightarrow \quad t \mapsto \left(x(t), \frac{dx}{dt}(t)\right) \text{ satisfies } \mathbf{E\text{-}L} .$$

The first line of $\mathbf{H}$ is automatically satisfied:

$$\frac{dx}{dt} = \frac{\partial H}{\partial \xi}(x, \xi) = L_{F_x^*}(\xi) = L_{F_x}^{-1}(\xi) \quad \Longleftrightarrow \quad \xi = L_{F_x}\left(\frac{dx}{dt}\right)$$

Claim. If $(x, \xi) = \mathcal{L}(x, v)$, then $\frac{\partial F}{\partial x}(x, v) = -\frac{\partial H}{\partial x}(x, \xi)$.

This follows from differentiating both sides of $H(x, \xi) = \xi \cdot v - F(x, v)$ with respect to x, where $\xi = L_{F_x}(v) = \xi(x, v)$.

$$\frac{\partial H}{\partial x} + \frac{\partial H}{\partial \xi}\underbrace{\frac{\partial \xi}{\partial x}}_{v} = \frac{\partial \xi}{\partial x} \cdot v - \frac{\partial F}{\partial x} .$$

Now the second line of $\mathbf{H}$ becomes

$$\underbrace{\frac{d}{dt}\frac{\partial F}{\partial v}(x, v)}_{\text{since } \xi = L_{F_x}(v)} = \frac{d\xi}{dt} = \underbrace{-\frac{\partial H}{\partial x}(x, \xi) = \frac{\partial F}{\partial x}(x, v)}_{\text{by the claim}} \quad \Longleftrightarrow \quad \mathbf{E\text{-}L} .$$

$$\square$$

Homework 15: Legendre Transform

This set of problems is adapted from [54].

1. Let $f : \mathbb{R} \to \mathbb{R}$ be a smooth function. f is called **strictly convex** if $f''(x) > 0$ for all $x \in \mathbb{R}$. Assuming that f is strictly convex, prove that the following four conditions are equivalent:

   (a) $f'(x) = 0$ for some point x_0,
   (b) f has a local minimum at some point x_0,
   (c) f has a unique (global) minimum at some point x_0,
   (d) $f(x) \to +\infty$ as $x \to \pm\infty$.

   The function f is **stable** if it satisfies one (and hence all) of these conditions. For what values of a is the function $e^x + ax$ stable? For those values of a for which it is not stable, what does the graph look like?

2. Let V be an n-dimensional vector space and $F : V \to \mathbb{R}$ a smooth function. The function F is said to be **strictly convex** if for every pair of elements $p, v \in V$, $v \neq 0$, the restriction of F to the line $\{p + xv \mid x \in \mathbb{R}\}$ is strictly convex. The **hessian** of F at p is the quadratic form

$$d^2 F_p : v \longmapsto \frac{d^2}{dx^2} F(p + xv)|_{x=0} \ .$$

   Show that F is strictly convex if and only if $d^2 F_p$ is positive definite for all $p \in V$. Prove the n-dimensional analogue of the result you proved in (1). Namely, assuming that F is strictly convex, show that the four following assertions are equivalent:

   (a) $dF_p = 0$ at some point p_0,
   (b) F has a local minimum at some point p_0,
   (c) F has a unique (global) minimum at some point p_0,
   (d) $F(p) \to +\infty$ as $p \to \infty$.

3. As in exercise 2, let V be an n-dimensional vector space and $F : V \to \mathbb{R}$ a smooth function. Since V is a vector space, there is a canonical identification $T_p^* V \simeq V^*$, for every $p \in V$. Therefore, we can define a map

$$L_F : V \longrightarrow V^* \qquad \textbf{(Legendre transform)}$$

   by setting

$$L_F(p) = dF_p \subset T_p^* V \simeq V^* \ .$$

   Show that, if F is strictly convex, then, for every point $p \in V$, L_F maps a neighborhood of p diffeomorphically onto a neighborhood of $L_F(p)$.

4. A strictly convex function $F : V \to \mathbb{R}$ is **stable** if it satisfies the four equivalent conditions of exercise 2. Given any strictly convex function F, we will denote

by S_F the set of $l \in V^*$ for which the function $F_l : V \to \mathbb{R}$, $p \mapsto F(p) - l(p)$, is stable. Prove that:

(a) The set S_F is open and convex.
(b) L_F maps V diffeomorphically onto S_F.
(c) If $\ell \in S_F$ and $p_0 = L_F^{-1}(\ell)$, then p_0 is the unique minimum point of the function F_ℓ.

Let $F^* : S_F \to \mathbb{R}$ be the function whose value at l is the quantity $-\min\limits_{p \in V} F_l(p)$.

Show that F^* is a smooth function.

The function F^* is called the **dual** of the function F.

5. Let F be a strictly convex function. F is said to have **quadratic growth at infinity** if there exists a positive-definite quadratic form Q on V and a constant K such that $F(p) \geq Q(p) - K$, for all p. Show that, if F has quadratic growth at infinity, then $S_F = V^*$ and hence L_F maps V diffeomorphically onto V^*.

6. Let $F : V \to \mathbb{R}$ be strictly convex and let $F^* : S_F \to \mathbb{R}$ be the dual function. Prove that for all $p \in V$ and all $\ell \in S_F$,

$$F(p) + F^*(\ell) \geq \ell(p) \qquad \textbf{(Young inequality)}.$$

7. On one hand we have $V \times V^* \simeq T^*V$, and on the other hand, since $V = V^{**}$, we have $V \times V^* \simeq V^* \times V \simeq T^*V^*$.

Let α_1 be the canonical 1-form on T^*V and α_2 be the canonical 1-form on T^*V^*. Via the identifications above, we can think of both of these forms as living on $V \times V^*$. Show that $\alpha_1 = d\beta - \alpha_2$, where $\beta : V \times V^* \to \mathbb{R}$ is the function $\beta(p, \ell) = \ell(p)$.

Conclude that the forms $\omega_1 = d\alpha_1$ and $\omega_2 = d\alpha_2$ satisfy $\omega_1 = -\omega_2$.

8. Let $F : V \to \mathbb{R}$ be strictly convex. Assume that F has quadratic growth at infinity so that $S_F = V^*$. Let Λ_F be the graph of the Legendre transform L_F. The graph Λ_F is a lagrangian submanifold of $V \times V^*$ with respect to the symplectic form ω_1; why? Hence, Λ_F is also lagrangian for ω_2.

Let $\mathrm{pr}_1 : \Lambda_F \to V$ and $\mathrm{pr}_2 : \Lambda_F \to V^*$ be the restrictions of the projection maps $V \times V^* \to V$ and $V \times V^* \to V^*$, and let $i : \Lambda_F \hookrightarrow V \times V^*$ be the inclusion map. Show that

$$i^*\alpha_1 = d(\mathrm{pr}_1)^*F \ .$$

Conclude that

$$i^*\alpha_2 = d(i^*\beta - (\mathrm{pr}_1)^*F) = d(\mathrm{pr}_2)^*F^* \ ,$$

and from this conclude that the inverse of the Legendre transform associated with F is the Legendre transform associated with F^*.

Part VIII
Moment Maps

The concept of a *moment map*[1] is a generalization of that of a hamiltonian function. The notion of a moment map associated to a group action on a symplectic manifold formalizes the Noether principle, which states that to every symmetry (such as a group action) in a mechanical system, there corresponds a conserved quantity.

---

[1] Souriau [95] invented the french name "application moment." In the US, East and West coasts could be distinguished by the choice of translation: *moment map* and *momentum map*, respectively. We will stick to the more economical version.

Chapter 21
Actions

21.1 One-Parameter Groups of Diffeomorphisms

Let M be a manifold and X a complete vector field on M. Let $\rho_t : M \to M$, $t \in \mathbb{R}$, be the family of diffeomorphisms generated by X. For each $p \in M$, $\rho_t(p)$, $t \in \mathbb{R}$, is by definition the unique integral curve of X passing through p at time 0, i.e., $\rho_t(p)$ satisfies

$$
\begin{cases}
\rho_0(p) = p \\[2mm]
\dfrac{d\rho_t(p)}{dt} = X(\rho_t(p)) .
\end{cases}
$$

Claim. We have that $\rho_t \circ \rho_s = \rho_{t+s}$.

Proof. Let $\rho_s(q) = p$. We need to show that $(\rho_t \circ \rho_s)(q) = \rho_{t+s}(q)$, for all $t \in \mathbb{R}$. Reparametrize as $\tilde{\rho}_t(q) := \rho_{t+s}(q)$. Then

$$
\begin{cases}
\tilde{\rho}_0(q) = \rho_s(q) = p \\[2mm]
\dfrac{d\tilde{\rho}_t(q)}{dt} = \dfrac{d\rho_{t+s}(q)}{dt} = X(\rho_{t+s}(q)) = X(\tilde{\rho}_t(q)) ,
\end{cases}
$$

i.e., $\tilde{\rho}_t(q)$ is an integral curve of X through p. By uniqueness we must have $\tilde{\rho}_t(q) = \rho_t(p)$, that is, $\rho_{t+s}(q) = \rho_t(\rho_s(q))$. $\qquad\square$

Consequence. We have that $\rho_t^{-1} = \rho_{-t}$.

In terms of the group $(\mathbb{R}, +)$ and the group $(\mathrm{Diff}(M), \circ)$ of all diffeomorphisms of M, these results can be summarized as:

Corollary 21.1. *The map* $\mathbb{R} \to \mathrm{Diff}(M)$, $t \mapsto \rho_t$, *is a group homomorphism.*

The family $\{\rho_t \mid t \in \mathbb{R}\}$ is then called a **one-parameter group of diffeomorphisms** of M and denoted

$$
\rho_t = \exp tX .
$$

21.2 Lie Groups

Definition 21.2. A *Lie group* is a manifold G equipped with a group structure where the group operations

$$G \times G \longrightarrow G \qquad \text{and} \qquad G \longrightarrow G$$
$$(a,b) \longmapsto a \cdot b \qquad\qquad\qquad a \longmapsto a^{-1}$$

are smooth maps.

Examples.

- $\mathbb{R}$ (with addition[1]).
- S^1 regarded as unit complex numbers with multiplication, represents rotations of the plane: $S^1 = \mathrm{U}(1) = \mathrm{SO}(2)$.
- $\mathrm{U}(n)$, unitary linear transformations of $\mathbb{C}^n$.
- $\mathrm{SU}(n)$, unitary linear transformations of $\mathbb{C}^n$ with $\det = 1$.
- $\mathrm{O}(n)$, orthogonal linear transformations of $\mathbb{R}^n$.
- $\mathrm{SO}(n)$, elements of $\mathrm{O}(n)$ with $\det = 1$.
- $\mathrm{GL}(V)$, invertible linear transformations of a vector space V.

$\diamondsuit$

Definition 21.3. A *representation* of a Lie group G on a vector space V is a group homomorphism $G \to \mathrm{GL}(V)$.

21.3 Smooth Actions

Let M be a manifold.

Definition 21.4. An *action* of a Lie group G on M is a group homomorphism

$$\psi : G \longrightarrow \mathrm{Diff}(M)$$
$$g \longmapsto \psi_g .$$

(We will only consider left actions where ψ is a homomorphism. A *right action* is defined with ψ being an anti-homomorphism.) The *evaluation map* associated with an action $\psi : G \to \mathrm{Diff}(M)$ is

$$\mathrm{ev}_\psi : M \times G \longrightarrow M$$
$$(p,g) \longmapsto \psi_g(p) .$$

The action ψ is **smooth** if ev_ψ is a smooth map.

---

[1] The operation will be omitted when it is clear from the context.

Example. If X is a complete vector field on M, then

$$\rho : \mathbb{R} \longrightarrow \text{Diff}(M)$$
$$t \longmapsto \rho_t = \exp tX$$

is a smooth action of $\mathbb{R}$ on M.   ◇

Every complete vector field gives rise to a smooth action of $\mathbb{R}$ on M. Conversely, every smooth action of $\mathbb{R}$ on M is defined by a complete vector field.

$$\{\text{complete vector fields on } M\} \overset{1-1}{\longleftrightarrow} \{\text{smooth actions of } \mathbb{R} \text{ on } M\}$$

$$X \longmapsto \exp tX$$

$$X_p = \left.\frac{d\psi_t(p)}{dt}\right|_{t=0} \longleftarrow \psi$$

21.4 Symplectic and Hamiltonian Actions

Let (M, ω) be a symplectic manifold, and G a Lie group. Let $\psi : G \longrightarrow \text{Diff}(M)$ be a (smooth) action.

Definition 21.5. The action ψ is a *symplectic action* if

$$\psi : G \longrightarrow \text{Sympl}(M, \omega) \subset \text{Diff}(M) ,$$

i.e., G "acts by symplectomorphisms."

$$\{\text{complete symplectic vector fields on } M\} \overset{1-1}{\longleftrightarrow} \{\text{symplectic actions of } \mathbb{R} \text{ on } M\}$$

Example. On $\mathbb{R}^{2n}$ with $\omega = \sum dx_i \wedge dy_i$, let $X = -\frac{\partial}{\partial y_1}$. The orbits of the action generated by X are lines parallel to the y_1-axis,

$$\{(x_1, y_1 - t, x_2, y_2, \ldots, x_n, y_n) \mid t \in \mathbb{R}\} .$$

Since $X = X_{x_1}$ is hamiltonian (with hamiltonian function $H = x_1$), this is actually an example of a *hamiltonian action* of $\mathbb{R}$.   ◇

Example. On S^2 with $\omega = d\theta \wedge dh$ (cylindrical coordinates), let $X = \frac{\partial}{\partial \theta}$. Each orbit is a horizontal circle (called a "parallel") $\{(\theta + t, h) \mid t \in \mathbb{R}\}$. Notice that all orbits of this $\mathbb{R}$-action close up after time 2π, so that this is an action of S^1:

$$\psi : S^1 \longrightarrow \text{Sympl}(S^2, \omega)$$
$$t \longmapsto \text{rotation by angle } t \text{ around } h\text{-axis} .$$

Since $X = X_h$ is hamiltonian (with hamiltonian function $H = h$), this is an example of a *hamiltonian action* of S^1.   ◇

Definition 21.6. A symplectic action ψ of S^1 or $\mathbb{R}$ on (M, ω) is **hamiltonian** if the vector field generated by ψ is hamiltonian. Equivalently, an action ψ of S^1 or $\mathbb{R}$ on (M, ω) is **hamiltonian** if there is $H : M \to \mathbb{R}$ with $dH = \iota_X \omega$, where X is the vector field generated by ψ.

What is a "hamiltonian action" of an arbitrary Lie group?

For the case where $G = \mathbb{T}^n = S^1 \times \cdots \times S^1$ is an n-torus, an action $\psi : G \to$ Sympl(M, ω) should be called *hamiltonian* when each restriction

$$\psi^i := \psi|_{\text{ith } S^1 \text{ factor}} : S^1 \longrightarrow \text{Sympl}(M, \omega)$$

is hamiltonian in the previous sense with hamiltonian function preserved by the action of the rest of G.

When G is not a product of S^1's or $\mathbb{R}$'s, the solution is to use an upgraded hamiltonian function, known as a *moment map*. Before its definition though (in Lecture 22), we need a little Lie theory.

21.5 Adjoint and Coadjoint Representations

Let G be a Lie group. Given $g \in G$ let

$$
\begin{aligned}
L_g : G &\longrightarrow G \\
a &\longmapsto g \cdot a
\end{aligned}
$$

be **left multiplication** by g. A vector field X on G is called **left-invariant** if $(L_g)_* X = X$ for every $g \in G$. (There are similar *right* notions.)

Let $\mathfrak{g}$ be the vector space of all left-invariant vector fields on G. Together with the Lie bracket $[\cdot, \cdot]$ of vector fields, $\mathfrak{g}$ forms a Lie algebra, called the **Lie algebra of the Lie group** G.

Exercise. Show that the map

$$
\begin{aligned}
\mathfrak{g} &\longrightarrow T_e G \\
X &\longmapsto X_e
\end{aligned}
$$

where e is the identity element in G, is an isomorphism of vector spaces.   ◇

Any Lie group G acts on itself by **conjugation**:

$$
\begin{aligned}
G &\longrightarrow \text{Diff}(G) \\
g &\longmapsto \psi_g, \qquad\qquad \psi_g(a) = g \cdot a \cdot g^{-1} .
\end{aligned}
$$

The derivative at the identity of

$$\psi_g : G \longrightarrow G$$
$$a \longmapsto g \cdot a \cdot g^{-1}$$

is an invertible linear map $\mathrm{Ad}_g : \mathfrak{g} \longrightarrow \mathfrak{g}$. Here we identify the Lie algebra $\mathfrak{g}$ with the tangent space T_eG. Letting g vary, we obtain the **adjoint representation** (or **adjoint action**) of G on $\mathfrak{g}$:

$$\mathrm{Ad} : G \longrightarrow GL(\mathfrak{g})$$
$$g \longmapsto \mathrm{Ad}_g \ .$$

Exercise. Check for matrix groups that

$$\frac{d}{dt}\mathrm{Ad}_{\exp tX}Y\bigg|_{t=0} = [X,Y] \ , \qquad \forall X,Y \in \mathfrak{g} \ .$$

Hint: For a matrix group G (i.e., a subgroup of $GL(n;\mathbb{R})$ for some n), we have

$$\mathrm{Ad}_g(Y) = gYg^{-1} \ , \qquad \forall g \in G \ , \ \forall Y \in \mathfrak{g}$$

and

$$[X,Y] = XY - YX \ , \qquad \forall X,Y \in \mathfrak{g} \ .$$

$\diamondsuit$

Let $\langle \cdot, \cdot \rangle$ be the natural pairing between $\mathfrak{g}^*$ and $\mathfrak{g}$:

$$\langle \cdot, \cdot \rangle : \mathfrak{g}^* \times \mathfrak{g} \longrightarrow \mathbb{R}$$
$$(\xi, X) \longmapsto \langle \xi, X \rangle = \xi(X) \ .$$

Given $\xi \in \mathfrak{g}^*$, we define $\mathrm{Ad}_g^*\xi$ by

$$\langle \mathrm{Ad}_g^*\xi, X \rangle = \langle \xi, \mathrm{Ad}_{g^{-1}}X \rangle \ , \qquad \text{for any } X \in \mathfrak{g} \ .$$

The collection of maps Ad_g^* forms the **coadjoint representation** (or **coadjoint action**) of G on $\mathfrak{g}^*$:

$$\mathrm{Ad}^* : G \longrightarrow GL(\mathfrak{g}^*)$$
$$g \longmapsto \mathrm{Ad}_g^* \ .$$

We take g^{-1} in the definition of $\mathrm{Ad}_g^*\xi$ in order to obtain a (left) representation, i.e., a group homomorphism, instead of a "right" representation, i.e., a group anti-homomorphism.

Exercise. Show that $\mathrm{Ad}_g \circ \mathrm{Ad}_h = \mathrm{Ad}_{gh}$ and $\mathrm{Ad}_g^* \circ \mathrm{Ad}_h^* = \mathrm{Ad}_{gh}^*$. $\diamondsuit$

Homework 16: Hermitian Matrices

Let $\mathcal{H}$ be the vector space of $n \times n$ complex hermitian matrices.

The unitary group $U(n)$ acts on $\mathcal{H}$ by conjugation: $\quad A \cdot \xi = A\xi A^{-1}$, for $A \in U(n)$, $\xi \in \mathcal{H}$.

For each $\lambda = (\lambda_1, \ldots, \lambda_n) \in \mathbb{R}^n$, let $\mathcal{H}_\lambda$ be the set of all $n \times n$ complex hermitian matrices whose spectrum is λ.

1. Show that the orbits of the $U(n)$-action are the manifolds $\mathcal{H}_\lambda$.

   For a fixed $\lambda \in \mathbb{R}^n$, what is the stabilizer of a point in $\mathcal{H}_\lambda$?

   Hint: If $\lambda_1, \ldots, \lambda_n$ are all distinct, the stabilizer of the diagonal matrix is the torus $\mathbb{T}^n$ of all diagonal unitary matrices.

2. Show that the symmetric bilinear form on $\mathcal{H}$, $(X, Y) \mapsto \text{trace } (XY)$, is nondegenerate.

   For $\xi \in \mathcal{H}$, define a skew-symmetric bilinear form ω_ξ on $\mathfrak{u}(n) = T_1 U(n) = i\mathcal{H}$ (space of skew-hermitian matrices) by

   $$\omega_\xi(X, Y) = i\,\text{trace } ([X, Y]\xi) , \qquad X, Y \in i\mathcal{H} .$$

   Check that $\omega_\xi(X, Y) = i\,\text{trace } (X(Y\xi - \xi Y))$ and $Y\xi - \xi Y \in \mathcal{H}$.

   Show that the kernel of ω_ξ is $K_\xi := \{Y \in \mathfrak{u}(n) \,|\, [Y, \xi] = 0\}$.

3. Show that K_ξ is the Lie algebra of the stabilizer of ξ.

   Hint: Differentiate the relation $A\xi A^{-1} = \xi$.

   Show that the ω_ξ's induce nondegenerate 2-forms on the orbits $\mathcal{H}_\lambda$.

   Show that these 2-forms are closed.

   Conclude that all the orbits $\mathcal{H}_\lambda$ are compact symplectic manifolds.

4. Describe the manifolds $\mathcal{H}_\lambda$.

   When all eigenvalues are equal, there is only one point in the orbit.

   Suppose that $\lambda_1 \neq \lambda_2 = \ldots = \lambda_n$. Then the eigenspace associated with λ_1 is a line, and the one associated with λ_2 is the orthogonal hyperplane. Show that there is a diffeomorphism $\mathcal{H}_\lambda \simeq \mathbb{CP}^{n-1}$. We have thus exhibited a lot of symplectic forms on $\mathbb{CP}^{n-1}$, on for each pair of distinct real numbers.

   What about the other cases?

   Hint: When the eigenvalues $\lambda_1 < \ldots < \lambda_n$ are all distinct, any element in $\mathcal{H}_\lambda$ defines a family of pairwise orthogonal lines in $\mathbb{C}^n$: its eigenspaces.

5. Show that, for any skew-hermitian matrix $X \in \mathfrak{u}(n)$, the vector field on $\mathcal{H}$ generated by $X \in \mathfrak{u}(n)$ for the $U(n)$-action by conjugation is $X_\xi^{\#} = [X, \xi]$.

Chapter 22
Hamiltonian Actions

22.1 Moment and Comoment Maps

Let

$$(M, \omega) \quad \text{be a symplectic manifold,}$$
$$G \quad \text{a Lie group, and}$$
$$\psi : G \to \text{Sympl}(M, \omega) \quad \text{a (smooth) symplectic action, i.e., a group homomorphism such that the evaluation map } \text{ev}_\psi(g, p) := \psi_g(p) \text{ is smooth.}$$

Case $G = \mathbb{R}$:

We have the following bijective correspondence:

$$\{\text{symplectic actions of } \mathbb{R} \text{ on } M\} \xleftrightarrow{1-1} \{\text{complete symplectic vector fields on } M\}$$

$$\psi \longmapsto X_p = \frac{d\psi_t(p)}{dt}$$

$$\psi = \exp tX \longleftarrow X$$

"flow of X" "vector field generated by ψ"

The action ψ is **hamiltonian** if there exists a function $H : M \to \mathbb{R}$ such that $dH = \iota_X \omega$ where X is the vector field on M generated by ψ.

Case $G = S^1$:

An action of S^1 is an action of $\mathbb{R}$ which is 2π-periodic: $\psi_{2\pi} = \psi_0$. The S^1-action is called **hamiltonian** if the underlying $\mathbb{R}$-action is hamiltonian.

General case:

Let

$$(M, \omega) \text{ be a symplectic manifold,}$$
$$G \quad \text{a Lie group,}$$
$$\mathfrak{g} \quad \text{the Lie algebra of } G,$$
$$\mathfrak{g}^* \text{ the dual vector space of } \mathfrak{g}, \text{ and}$$

$$\psi : G \longrightarrow \text{Sympl}(M, \omega) \text{ a symplectic action.}$$

Definition 22.1. The action ψ is a **hamiltonian action** if there exists a map

$$\mu : M \longrightarrow \mathfrak{g}^*$$

satisfying:

1. For each $X \in \mathfrak{g}$, let

   - $\mu^X : M \to \mathbb{R}$, $\mu^X(p) := \langle \mu(p), X \rangle$, be the component of μ along X,
   - $X^\#$ be the vector field on M generated by the one-parameter subgroup $\{\exp tX \mid t \in \mathbb{R}\} \subseteq G$.

   Then

   $$d\mu^X = \iota_{X^\#} \omega$$

   i.e., μ^X is a hamiltonian function for the vector field $X^\#$.

2. μ is *equivariant* with respect to the given action ψ of G on M and the coadjoint action Ad^* of G on $\mathfrak{g}^*$:

   $$\mu \circ \psi_g = \text{Ad}^*_g \circ \mu , \qquad \text{for all } g \in G .$$

The vector (M, ω, G, μ) is then called a **hamiltonian G-space** and μ is a **moment map**.

For connected Lie groups, hamiltonian actions can be equivalently defined in terms of a **comoment map**

$$\mu^* : \mathfrak{g} \longrightarrow C^\infty(M) ,$$

with the two conditions rephrased as:

1. $\mu^*(X) := \mu^X$ is a hamiltonian function for the vector field $X^\#$,
2. μ^* is a Lie algebra homomorphism:

   $$\mu^*[X, Y] = \{\mu^*(X), \mu^*(Y)\}$$

where $\{\cdot, \cdot\}$ is the Poisson bracket on $C^\infty(M)$.

These definitions match the previous ones for the cases $G = \mathbb{R}, S^1$, torus, where equivariance becomes invariance since the coadjoint action is trivial.

Case $G = S^1$ (or $\mathbb{R}$):

Here $\mathfrak{g} \simeq \mathbb{R}$, $\mathfrak{g}^* \simeq \mathbb{R}$. A moment map $\mu : M \longrightarrow \mathbb{R}$ satisfies:

1. For the generator $X = 1$ of $\mathfrak{g}$, we have $\mu^X(p) = \mu(p) \cdot 1$, i.e., $\mu^X = \mu$, and $X^\#$ is the standard vector field on M generated by S^1. Then $d\mu = \iota_{X^\#}\omega$.
2. μ is invariant: $\mathcal{L}_{X^\#}\mu = \iota_{X^\#}d\mu = 0$.

Case $G = \mathbb{T}^n = n$-torus:

Here $\mathfrak{g} \simeq \mathbb{R}^n$, $\mathfrak{g}^* \simeq \mathbb{R}^n$. A moment map $\mu : M \longrightarrow \mathbb{R}^n$ satisfies:

1. For each basis vector X_i of $\mathbb{R}^n$, μ^{X_i} is a hamiltonian function for $X_i^\#$.
2. μ is invariant.

22.2 Orbit Spaces

Let $\psi : G \to \mathrm{Diff}(M)$ be any action.

Definition 22.2. The *orbit* of G through $p \in M$ is $\{\psi_g(p) \mid g \in G\}$.
The *stabilizer* (or *isotropy*) of $p \in M$ is the subgroup $G_p := \{g \in G \mid \psi_g(p) = p\}$.

Exercise. If q is in the orbit of p, then G_q and G_p are conjugate subgroups. $\diamondsuit$

Definition 22.3. We say that the action of G on M is ...

- *transitive* if there is just one orbit,
- *free* if all stabilizers are trivial $\{e\}$,
- *locally free* if all stabilizers are discrete.

Let $\sim$ be the orbit equivalence relation; for $p, q \in M$,

$$p \sim q \quad \Longleftrightarrow \quad p \text{ and } q \text{ are on the same orbit.}$$

The space of orbits $M/\sim = M/G$ is called the **orbit space**. Let

$$\pi : M \longrightarrow M/G$$
$$p \longmapsto \text{orbit through } p$$

be the **point-orbit projection**.

Topology of the orbit space:

We equip M/G with the weakest topology for which π is continuous, i.e., $\mathcal{U} \subseteq M/G$ is open if and only if $\pi^{-1}(\mathcal{U})$ is open in M. This is called the **quotient topology**. This topology can be "bad." For instance:

Example. Let $G = \mathbb{R}$ act on $M = \mathbb{R}$ by

$$t \longmapsto \psi_t = \text{multiplication by } e^t .$$

There are three orbits $\mathbb{R}^+$, $\mathbb{R}^-$ and $\{0\}$. The point in the three-point orbit space corresponding to the orbit $\{0\}$ is not open, so the orbit space with the quotient topology is *not* Hausdorff. $\diamondsuit$

Example. Let $G = \mathbb{C}\backslash\{0\}$ act on $M = \mathbb{C}^n$ by

$$\lambda \longmapsto \psi_\lambda = \text{multiplication by } \lambda .$$

The orbits are the punctured complex lines (through non-zero vectors $z \in \mathbb{C}^n$), plus one "unstable" orbit through 0, which has a single point. The orbit space is

$$M/G = \mathbb{CP}^{n-1} \sqcup \{\text{point}\} .$$

The quotient topology restricts to the usual topology on $\mathbb{CP}^{n-1}$. The only open set containing $\{\text{point}\}$ in the quotient topology is the full space. Again the quotient topology in M/G is *not* Hausdorff.

However, it suffices to remove 0 from $\mathbb{C}^n$ to obtain a Hausdorff orbit space: $\mathbb{CP}^{n-1}$. Then there is also a compact (yet not complex) description of the orbit space by taking only unit vectors:

$$\mathbb{CP}^{n-1} = \left(\mathbb{C}^n\backslash\{0\} \right)\big/\left(\mathbb{C}\backslash\{0\} \right) = S^{2n-1}/S^1 .$$

22.3 Preview of Reduction

Let $\omega = \frac{i}{2}\sum dz_i \wedge d\bar{z}_i = \sum dx_i \wedge dy_i = \sum r_i dr_i \wedge d\theta_i$ be the standard symplectic form on $\mathbb{C}^n$. Consider the following S^1-action on $(\mathbb{C}^n, \omega)$:

$$t \in S^1 \longmapsto \psi_t = \text{multiplication by } t .$$

The action ψ is hamiltonian with moment map

$$\mu : \mathbb{C}^n \longrightarrow \mathbb{R}$$
$$z \longmapsto -\frac{|z|^2}{2} + \text{constant}$$

since

$$d\mu = -\tfrac{1}{2}d(\textstyle\sum r_i^2)$$

$$X^\# = \frac{\partial}{\partial \theta_1} + \frac{\partial}{\partial \theta_2} + \cdots + \frac{\partial}{\partial \theta_n}$$

$$\iota_{X^\#}\omega = -\sum r_i dr_i = -\tfrac{1}{2}\sum dr_i^2 \ .$$

If we choose the constant to be $\frac{1}{2}$, then $\mu^{-1}(0) = S^{2n-1}$ is the unit sphere. The orbit space of the zero level of the moment map is

$$\mu^{-1}(0)/S^1 = S^{2n-1}/S^1 = \mathbb{CP}^{n-1} \ .$$

$\mathbb{CP}^{n-1}$ is thus called a **reduced space**. Notice also that the image of the moment map is half-space.

These particular observations are related to major theorems:

Under assumptions (explained in Lectures 23-29),

- [Marsden-Weinstein-Meyer] reduced spaces are symplectic manifolds;
- [Atiyah-Guillemin-Sternberg] the image of the moment map is a convex polytope;
- [Delzant] hamiltonian $\mathbb{T}^n$-spaces are classified by the image of the moment map.

22.4 Classical Examples

Example. Let $G = SO(3) = \{A \in GL(3;\mathbb{R}) \mid A\ {}^tA = \mathrm{Id}$ and $\det A = 1\}$. Then $\mathfrak{g} = \{A \in \mathfrak{gl}(3;\mathbb{R}) \mid A + A^t = 0\}$ is the space of 3×3 skew-symmetric matrices and can be identified with $\mathbb{R}^3$. The Lie bracket on $\mathfrak{g}$ can be identified with the exterior product via

$$A = \begin{bmatrix} 0 & -a_3 & a_2 \\ a_3 & 0 & -a_1 \\ -a_2 & a_1 & 0 \end{bmatrix} \longmapsto \overrightarrow{a} = (a_1, a_2, a_3)$$

$$[A, B] = AB - BA \longmapsto \overrightarrow{a} \times \overrightarrow{b} \ .$$

Exercise. Under the identifications $\mathfrak{g}, \mathfrak{g}^* \simeq \mathbb{R}^3$, the adjoint and coadjoint actions are the usual $SO(3)$-action on $\mathbb{R}^3$ by rotations. $\diamondsuit$

Therefore, the coadjoint orbits are the spheres in $\mathbb{R}^3$ centered at the origin. Homework 17 shows that coadjoint orbits are symplectic. $\diamondsuit$

The name "moment map" comes from being the generalization of linear and angular momenta in classical mechanics.

Translation: Consider $\mathbb{R}^6$ with coordinates $x_1, x_2, x_3, y_1, y_2, y_3$ and symplectic form $\omega = \sum dx_i \wedge dy_i$. Let $\mathbb{R}^3$ act on $\mathbb{R}^6$ by translations:

$$\vec{a} \in \mathbb{R}^3 \longmapsto \psi_{\vec{a}} \in \mathrm{Sympl}(\mathbb{R}^6, \omega)$$

$$\psi_{\vec{a}}(\vec{x}, \vec{y}) = (\vec{x} + \vec{a}, \vec{y}) \,.$$

Then $X^{\#} = a_1 \frac{\partial}{\partial x_1} + a_2 \frac{\partial}{\partial x_2} + a_3 \frac{\partial}{\partial x_3}$ for $X = \vec{a}$, and

$$\mu : \mathbb{R}^6 \longrightarrow \mathbb{R}^3 \,, \quad \mu(\vec{x}, \vec{y}) = \vec{y}$$

is a moment map, with

$$\mu^{\vec{a}}(\vec{x}, \vec{y}) = \langle \mu(\vec{x}, \vec{y}), \vec{a} \rangle = \vec{y} \cdot \vec{a} \,.$$

Classically, $\vec{y}$ is called the **momentum vector** corresponding to the **position vector** $\vec{x}$, and the map μ is called the **linear momentum**.

Rotation: The $\mathrm{SO}(3)$-action on $\mathbb{R}^3$ by rotations lifts to a symplectic action ψ on the cotangent bundle $\mathbb{R}^6$. The infinitesimal version of this action is

$$\vec{a} \in \mathbb{R}^3 \longmapsto d\psi(\vec{a}) \in \chi^{\mathrm{sympl}}(\mathbb{R}^6)$$

$$d\psi(\vec{a})(\vec{x}, \vec{y}) = (\vec{a} \times \vec{x}, \vec{a} \times \vec{y}) \,.$$

Then

$$\mu : \mathbb{R}^6 \longrightarrow \mathbb{R}^3 \,, \quad \mu(\vec{x}, \vec{y}) = \vec{x} \times \vec{y}$$

is a moment map, with

$$\mu^{\vec{a}}(\vec{x}, \vec{y}) = \langle \mu(\vec{x}, \vec{y}), \vec{a} \rangle = (\vec{x} \times \vec{y}) \cdot \vec{a} \,.$$

The map μ is called the **angular momentum**.

Homework 17: Coadjoint Orbits

Let G be a Lie group, $\mathfrak{g}$ its Lie algebra and $\mathfrak{g}^*$ the dual vector space of $\mathfrak{g}$.

1. Let $\mathfrak{g}X^{\#}$ be the vector field generated by $X \in \mathfrak{g}$ for the adjoint representation of G on $\mathfrak{g}$. Show that

$$\mathfrak{g}X_Y^{\#} = [X,Y] \qquad \forall Y \in \mathfrak{g}.$$

2. Let $X^{\#}$ be the vector field generated by $X \in \mathfrak{g}$ for the coadjoint representation of G on $\mathfrak{g}^*$. Show that

$$\langle X_\xi^{\#}, Y \rangle = \langle \xi, [Y,X] \rangle \qquad \forall Y \in \mathfrak{g}.$$

3. For any $\xi \in \mathfrak{g}^*$, define a skew-symmetric bilinear form on $\mathfrak{g}$ by

$$\omega_\xi(X,Y) := \langle \xi, [X,Y] \rangle.$$

   Show that the kernel of ω_ξ is the Lie algebra $\mathfrak{g}_\xi$ of the stabilizer of ξ for the coadjoint representation.
4. Show that ω_ξ defines a nondegenerate 2-form on the tangent space at ξ to the coadjoint orbit through ξ.
5. Show that ω_ξ defines a closed 2-form on the orbit of ξ in $\mathfrak{g}^*$.

   Hint: The tangent space to the orbit being generated by the vector fields $X^{\#}$, this is a consequence of the Jacobi identity in $\mathfrak{g}$.

   This **canonical symplectic form** on the coadjoint orbits in $\mathfrak{g}^*$ is also known as the **Lie-Poisson** or **Kostant-Kirillov symplectic structure**.
6. The Lie algebra structure of $\mathfrak{g}$ defines a canonical Poisson structure on $\mathfrak{g}^*$:

$$\{f,g\}(\xi) := \langle \xi, [df_\xi, dg_\xi] \rangle$$

   for $f,g \in C^\infty(\mathfrak{g}^*)$ and $\xi \in \mathfrak{g}^*$. Notice that $df_\xi : T_\xi \mathfrak{g}^* \simeq \mathfrak{g}^* \to \mathbb{R}$ is identified with an element of $\mathfrak{g} \simeq \mathfrak{g}^{**}$.
   Check that $\{\cdot,\cdot\}$ satisfies the Leibniz rule:

$$\{f,gh\} = g\{f,h\} + h\{f,g\}.$$

7. Show that the **jacobiator**

$$J(f,g,h) := \{\{f,g\},h\} + \{\{g,h\},f\} + \{\{h,f\},g\}$$

   is a trivector field, i.e., J is a skew-symmetric trilinear map $C^\infty(\mathfrak{g}^*) \times C^\infty(\mathfrak{g}^*) \times C^\infty(\mathfrak{g}^*) \to C^\infty(\mathfrak{g}^*)$, which is a derivation in each argument.

   Hint: Being a derivation amounts to the Leibniz rule from exercise 6.

8. Show that $J \equiv 0$, i.e., $\{\cdot,\cdot\}$ satisfies the Jacobi identity.

   Hint: Follows from the Jacobi identity for $[\cdot,\cdot]$ in $\mathfrak{g}$. It is enough to check on coordinate functions.

Part IX
Symplectic Reduction

The phase space of a system of n particles is the space parametrizing the position and momenta of the particles. The mathematical model for the phase space is a symplectic manifold. Classical physicists realized that, whenever there is a symmetry group of dimension k acting on a mechanical system, then the number of degrees of freedom for the position and momenta of the particles may be reduced by $2k$. Symplectic reduction formulates this feature mathematically.

Chapter 23
The Marsden-Weinstein-Meyer Theorem

23.1 Statement

Theorem 23.1. *(Marsden-Weinstein-Meyer [77, 85])* Let (M, ω, G, μ) be a hamiltonian G-space for a compact Lie group G. Let $i : \mu^{-1}(0) \hookrightarrow M$ be the inclusion map. Assume that G acts freely on $\mu^{-1}(0)$. Then

- *the orbit space $M_{\text{red}} = \mu^{-1}(0)/G$ is a manifold,*
- *$\pi : \mu^{-1}(0) \to M_{\text{red}}$ is a principal G-bundle, and*
- *there is a symplectic form ω_{red} on M_{red} satisfying $i^*\omega = \pi^*\omega_{\text{red}}$.*

Definition 23.2. The pair $(M_{\text{red}}, \omega_{\text{red}})$ is called the *reduction* of (M, ω) with respect to G, μ, or the *reduced space*, or the *symplectic quotient*, or the *Marsden-Weinstein-Meyer quotient*, etc.

Low-brow proof for the case $G = S^1$ and $\dim M = 4$.

In this case the moment map is $\mu : M \to \mathbb{R}$. Let $p \in \mu^{-1}(0)$. Choose local coordinates:

- θ along the orbit through p,
- μ given by the moment map, and
- η_1, η_2 pullback of coordinates on $\mu^{-1}(0)/S^1$.

Then the symplectic form can be written

$$\omega = A\, d\theta \wedge d\mu + B_j\, d\theta \wedge d\eta_j + C_j\, d\mu \wedge d\eta_j + D\, d\eta_1 \wedge d\eta_2 .$$

Since $d\mu = \iota\left(\frac{\partial}{\partial\theta}\right)\omega$, we must have $A - 1$, $B_j - 0$. Hence,

$$\omega = d\theta \wedge d\mu + C_j\, d\mu \wedge d\eta_j + D\, d\eta_1 \wedge d\eta_2 .$$

Since ω is symplectic, we must have $D \neq 0$. Therefore, $i^*\omega = D\, d\eta_1 \wedge d\eta_2$ is the pullback of a symplectic form on M_{red}. □

The actual proof of the Marsden-Weinstein-Meyer theorem requires the following ingredients.

23.2 Ingredients

1. Let $\mathfrak{g}_p$ be the Lie algebra of the stabilizer of $p \in M$. Then $d\mu_p : T_pM \to \mathfrak{g}^*$ has

$$\ker d\mu_p = (T_p\mathcal{O}_p)^{\omega_p}$$
$$\operatorname{im} d\mu_p = \mathfrak{g}_p^0$$

where $\mathcal{O}_p$ is the G-orbit through p, and $\mathfrak{g}_p^0 = \{\xi \in \mathfrak{g}^* \mid \langle \xi, X \rangle = 0, \forall X \in \mathfrak{g}_p\}$ is the annihilator of $\mathfrak{g}_p$.

Proof. Stare at the expression $\omega_p(X_p^\#, v) = \langle d\mu_p(v), X \rangle$, for all $v \in T_pM$ and all $X \in \mathfrak{g}$, and count dimensions. $\square$

Consequences:

- The action is locally free at p
  $\iff \mathfrak{g}_p = \{0\}$
  $\iff d\mu_p$ is surjective
  $\iff p$ is a regular point of μ.
- G acts freely on $\mu^{-1}(0)$
  $\implies 0$ is a regular value of μ
  $\implies \mu^{-1}(0)$ is a closed submanifold of M
  of codimension equal to $\dim G$.
- G acts freely on $\mu^{-1}(0)$
  $\implies T_p\mu^{-1}(0) = \ker d\mu_p$ (for $p \in \mu^{-1}(0)$)
  $\implies T_p\mu^{-1}(0)$ and $T_p\mathcal{O}_p$ are symplectic orthocomplements in T_pM.
  In particular, the tangent space to the orbit through $p \in \mu^{-1}(0)$ is an isotropic subspace of T_pM. Hence, orbits in $\mu^{-1}(0)$ are isotropic.

Since any tangent vector to the orbit is the value of a vector field generated by the group, we can confirm that orbits are isotropic directly by computing, for any $X, Y \in \mathfrak{g}$ and any $p \in \mu^{-1}(0)$,

$$\omega_p(X_p^\#, Y_p^\#) = \text{hamiltonian function for } [Y^\#, X^\#] \text{ at } p$$
$$= \text{hamiltonian function for } [Y, X]^\# \text{ at } p$$
$$= \mu^{[Y,X]}(p) = 0 \,.$$

2. **Lemma 23.3.** *Let (V, ω) be a symplectic vector space. Suppose that I is an isotropic subspace, that is, $\omega|_I \equiv 0$. Then ω induces a canonical symplectic form Ω on I^ω / I.*

Proof. Let $u, v \in I^\omega$, and $[u], [v] \in I^\omega / I$. Define $\Omega([u], [v]) = \omega(u, v)$.

- Ω is well-defined:

$$\omega(u+i, v+j) = \omega(u,v) + \underbrace{\omega(u,j)}_{0} + \underbrace{\omega(i,v)}_{0} + \underbrace{\omega(i,j)}_{0}, \quad \forall i, j \in I \,.$$

- Ω is nondegenerate:
  Suppose that $u \in I^\omega$ has $\omega(u, v) = 0$, for all $v \in I^\omega$.
  Then $u \in (I^\omega)^\omega = I$, i.e., $[u] = 0$.

  $\square$

3. **Theorem 23.4.** *If a compact Lie group G acts freely on a manifold M, then M/G is a manifold and the map $\pi : M \to M/G$ is a principal G-bundle.*

   Proof. We will first show that, for any $p \in M$, the G-orbit through p is a compact embedded submanifold of M diffeomorphic to G.

   Since the action is smooth, the evaluation map $\mathrm{ev} : G \times M \to M$, $\mathrm{ev}(g, p) = g \cdot p$, is smooth. Let $\mathrm{ev}_p : G \to M$ be defined by $\mathrm{ev}_p(g) = g \cdot p$. The map ev_p provides the embedding we seek:

   The image of ev_p is the G-orbit through p. Injectivity of ev_p follows from the action of G being free. The map ev_p is proper because, if A is a compact, hence closed, subset of M, then its inverse image $(\mathrm{ev}_p)^{-1}(A)$, being a closed subset of the compact Lie group G, is also compact. It remains to show that ev_p is an immersion. For $X \in \mathfrak{g} \simeq T_e G$, we have

   $$d(\mathrm{ev}_p)_e(X) = 0 \iff X_p^\# = 0 \iff X = 0,$$

   as the action is free. We conclude that $d(\mathrm{ev}_p)_e$ is injective. At any other point $g \in G$, for $X \in T_g G$, we have

   $$d(\mathrm{ev}_p)_g(X) = 0 \iff d(\mathrm{ev}_p \circ R_g)_e \circ (dR_{g^{-1}})_g(X) = 0,$$

   where $R_g : G \to G$ is right multiplication by g. But $\mathrm{ev}_p \circ R_g = \mathrm{ev}_{g \cdot p}$ has an injective differential at e, and $(dR_{g^{-1}})_g$ is an isomorphism. It follows that $d(\mathrm{ev}_p)_g$ is always injective.

   Exercise. Show that, even if the action is not free, the G-orbit through p is a compact embedded submanifold of M. In that case, the orbit is diffeomorphic to the quotient of G by the isotropy of p: $\mathcal{O}_p \simeq G/G_p$. $\diamond$

   Let S be a transverse section to $\mathcal{O}_p$ at p; this is called a **slice**. Choose a coordinate system $x_1, \ldots, x_n$ centered at p such that

   $$\begin{aligned} \mathcal{O}_p \simeq G &: x_1 &&= \ldots = x_k = 0 \\ S \quad\; &: x_{k+1} &&= \ldots = x_n = 0 \,. \end{aligned}$$

   Let $S_\varepsilon = S \cap B_\varepsilon(0, \mathbb{R}^n)$ where $B_\varepsilon(0, \mathbb{R}^n)$ is the ball of radius ε centered at 0 in $\mathbb{R}^n$. Let $\eta : G \times S \to M$, $\eta(g, s) = g \cdot s$. Apply the following equivariant tubular neighborhood theorem.

   Theorem 23.5. (Slice Theorem) *Let G be a compact Lie group acting on a manifold M such that G acts freely at $p \in M$. For sufficiently small ε, $\eta : G \times S_\varepsilon \to M$ maps $G \times S_\varepsilon$ diffeomorphically onto a G-invariant neighborhood $\mathcal{U}$ of the G-orbit through p.*

The proof of this slice theorem is sketched further below.

Corollary 23.6. *If the action of G is free at p, then the action is free on $\mathcal{U}$.*

Corollary 23.7. *The set of points where G acts freely is open.*

Corollary 23.8. *The set $G \times S_\varepsilon \simeq \mathcal{U}$ is G-invariant. Hence, the quotient $\mathcal{U}/G \simeq S_\varepsilon$ is smooth.*

Conclusion of the proof that M/G is a manifold and $\pi : M \to M/G$ is a smooth fiber map.

For $p \in M$, let $q = \pi(p) \in M/G$. Choose a G-invariant neighborhood $\mathcal{U}$ of p as in the slice theorem: $\mathcal{U} \simeq G \times S$ (where $S = S_\varepsilon$ for an appropriate ε). Then $\pi(\mathcal{U}) = \mathcal{U}/G =: \mathcal{V}$ is an open neighborhood of q in M/G. By the slice theorem, $S \xrightarrow{\simeq} \mathcal{V}$ is a homeomorphism. We will use such neighborhoods $\mathcal{V}$ as charts on M/G. To show that the transition functions associated with these charts are smooth, consider two G-invariant open sets $\mathcal{U}_1, \mathcal{U}_2$ in M and corresponding slices S_1, S_2 of the G-action. Then $S_{12} = S_1 \cap \mathcal{U}_2$, $S_{21} = S_2 \cap \mathcal{U}_1$ are both slices for the G-action on $\mathcal{U}_1 \cap \mathcal{U}_2$. To compute the transition map $S_{12} \to S_{21}$, consider the diagram

$$S_{12} \xrightarrow{\simeq} \mathrm{id} \times S_{12} \hookrightarrow G \times S_{12}$$
$$\searrow{\scriptstyle\simeq}$$
$$\mathcal{U}_1 \cap \mathcal{U}_2 \; .$$
$$\nearrow{\scriptstyle\simeq}$$
$$S_{21} \xrightarrow{\simeq} \mathrm{id} \times S_{21} \hookrightarrow G \times S_{21}$$

Then the composition

$$S_{12} \hookrightarrow \mathcal{U}_1 \cap \mathcal{U}_2 \xrightarrow{\simeq} G \times S_{21} \xrightarrow{pr} S_{21}$$

is smooth.

Finally, we need to show that $\pi : M \to M/G$ is a smooth fiber map. For $p \in M$, $q = \pi(p)$, choose a G-invariant neighborhood $\mathcal{U}$ of the G-orbit through p of the form $\eta : G \times S \xrightarrow{\simeq} \mathcal{U}$. Then $\mathcal{V} = \mathcal{U}/G \simeq S$ is the corresponding neighborhood of q in M/G:

$$
\begin{array}{ccccc}
M \supseteq & \mathcal{U} & \overset{\eta}{\underset{\simeq}{=}} & G \times S & \simeq G \times \mathcal{V} \\
 & \downarrow \pi & & & \downarrow \\
M/G \supseteq & \mathcal{V} & = & & \mathcal{V}
\end{array}
$$

Since the projection on the right is smooth, π is smooth.

Exercise. Check that the transition functions for the bundle defined by π are smooth. $\diamondsuit$

$\square$

Sketch for the proof of the slice theorem. We need to show that, for ε sufficiently small, $\eta : G \times S_\varepsilon \to \mathcal{U}$ is a diffeomorphism where $\mathcal{U} \subseteq M$ is a G-invariant neighborhood of the G-orbit through p. Show that:

(a) $d\eta_{(\mathrm{id},p)}$ is bijective.
(b) Let G act on $G \times S$ by the product of its left action on G and trivial action on S. Then $\eta : G \times S \to M$ is G-equivariant.
(c) $d\eta$ is bijective at all points of $G \times \{p\}$. This follows from (a) and (b).
(d) The set $G \times \{p\}$ is compact, and $\eta : G \times S \to M$ is injective on $G \times \{p\}$ with $d\eta$ bijective at all these points. By the implicit function theorem, there is a neighborhood $\mathcal{U}_0$ of $G \times \{p\}$ in $G \times S$ such that η maps $\mathcal{U}_0$ diffeomorphically onto a neighborhood $\mathcal{U}$ of the G-orbit through p.
(e) The sets $G \times S_\varepsilon$, varying ε, form a neighborhood base for $G \times \{p\}$ in $G \times S$. So in (d) we may take $\mathcal{U}_0 = G \times S_\varepsilon$.

$\square$

23.3 Proof of the Marsden-Weinstein-Meyer Theorem

Since

$$G \text{ acts freely on } \mu^{-1}(0) \implies d\mu_p \text{ is surjective for all } p \in \mu^{-1}(0)$$
$$\implies 0 \text{ is a regular value}$$
$$\implies \mu^{-1}(0) \text{ is a submanifold of codimension} = \dim G$$

for the first two parts of the Marsden-Weinstein-Meyer theorem it is enough to apply the third ingredient from Section 23.2 to the free action of G on $\mu^{-1}(0)$.

At $p \in \mu^{-1}(0)$ the tangent space to the orbit $T_p\mathcal{O}_p$ is an isotropic subspace of the symplectic vector space (T_pM, ω_p), i.e., $T_p\mathcal{O}_p \subseteq (T_p\mathcal{O}_p)^\omega$.

$$(T_p\mathcal{O}_p)^\omega = \ker d\mu_p = T_p\mu^{-1}(0) .$$

The lemma (second ingredient) gives a canonical symplectic structure on the quotient $T_p\mu^{-1}(0)/T_p\mathcal{O}_p$. The point $[p] \in M_{\mathrm{red}} = \mu^{-1}(0)/G$ has tangent space $T_{[p]}M_{\mathrm{red}} \simeq T_p\mu^{-1}(0)/T_p\mathcal{O}_p$. Thus the lemma defines a nondegenerate 2-form ω_{red} on M_{red}. This is well-defined because ω is G-invariant.

By construction $i^*\omega = \pi^*\omega_{\mathrm{red}}$ where

$$\mu^{-1}(0) \overset{i}{\hookrightarrow} M$$
$$\downarrow \pi$$
$$M_{\mathrm{red}}$$

Hence, $\pi^*d\omega_{\mathrm{red}} = d\pi^*\omega_{\mathrm{red}} = di^*\omega = i^*d\omega = 0$. The closedness of ω_{red} follows from the injectivity of π^*. $\square$

Remark. Suppose that another Lie group H acts on (M, ω) in a hamiltonian way with moment map $\phi : M \to \mathfrak{h}^*$. If the H-action commutes with the G-action, and if ϕ is G-invariant, then M_{red} inherits a hamiltonian action of H, with moment map $\phi_{\mathrm{red}} : M_{\mathrm{red}} \to \mathfrak{h}^*$ satisfying $\phi_{\mathrm{red}} \circ \pi = \phi \circ i$. $\diamond$

Chapter 24
Reduction

24.1 Noether Principle

Let (M, ω, G, μ) be a hamiltonian G-space.

Theorem 24.1. *(Noether)* *A function $f : M \to \mathbb{R}$ is G-invariant if and only if μ is constant on the trajectories of the hamiltonian vector field of f.*

Proof. Let v_f be the hamiltonian vector field of f. Let $X \in \mathfrak{g}$ and $\mu^X = \langle \mu, X \rangle :$ $M \to \mathbb{R}$. We have

$$\mathcal{L}_{v_f} \mu^X = \iota_{v_f} d\mu^X = \iota_{v_f} \iota_{X^\#} \omega$$
$$= -\iota_{X^\#} \iota_{v_f} \omega = -\iota_{X^\#} df$$
$$= -\mathcal{L}_{X^\#} f = 0$$

because f is G-invariant. $\qquad\qquad\square$

Definition 24.2. A G-invariant function $f : M \to \mathbb{R}$ is called an **integral of motion** of (M, ω, G, μ). If μ is constant on the trajectories of a hamiltonian vector field v_f, then the corresponding one-parameter group of diffeomorphisms $\{\exp t v_f \mid t \in \mathbb{R}\}$ is called a **symmetry** of (M, ω, G, μ).

The **Noether principle** asserts that there is a one-to-one correspondence between symmetries and integrals of motion.

24.2 Elementary Theory of Reduction

Finding a symmetry for a $2n$-dimensional mechanical problem may reduce it to a $(2n-2)$-dimensional problem as follows: an integral of motion f for a $2n$-dimensional hamiltonian system (M, ω, H) may enable us to understand the trajectories of this system in terms of the trajectories of a $(2n-2)$-dimensional hamiltonian system $(M_{red}, \omega_{red}, H_{red})$. To make this precise, we will describe this process locally. Suppose that $\mathcal{U}$ is an open set in M with Darboux coordinates $x_1, \ldots, x_n, \xi_1, \ldots, \xi_n$ such that $f = \xi_n$ for this chart, and write H in these coordinates: $H = H(x_1, \ldots, x_n, \xi_1, \ldots, \xi_n)$. Then

$$\xi_n \text{ is an integral of motion} \implies \begin{cases} \text{the trajectories of } v_H \text{ lie on the} \\ \text{hyperplane } \xi_n = \text{constant} \\ \{\xi_n, H\} = 0 = -\frac{\partial H}{\partial x_n} \\ \qquad \implies H = H(x_1, \ldots, x_{n-1}, \xi_1, \ldots, \xi_n). \end{cases}$$

If we set $\xi_n = c$, the motion of the system on this hyperplane is described by the following Hamilton equations:

$$\begin{cases} \dfrac{dx_1}{dt} = \dfrac{\partial H}{\partial \xi_1}(x_1, \ldots, x_{n-1}, \xi_1, \ldots, \xi_{n-1}, c) \\[2mm] \quad \vdots \\[2mm] \dfrac{dx_{n-1}}{dt} = \dfrac{\partial H}{\partial \xi_{n-1}}(x_1, \ldots, x_{n-1}, \xi_1, \ldots, \xi_{n-1}, c) \\[2mm] \dfrac{d\xi_1}{dt} = -\dfrac{\partial H}{\partial x_1}(x_1, \ldots, x_{n-1}, \xi_1, \ldots, \xi_{n-1}, c) \\[2mm] \quad \vdots \\[2mm] \dfrac{d\xi_{n-1}}{dt} = -\dfrac{\partial H}{\partial x_{n-1}}(x_1, \ldots, x_{n-1}, \xi_1, \ldots, \xi_{n-1}, c) \end{cases}$$

$$\dfrac{dx_n}{dt} = \dfrac{\partial H}{\partial \xi_n}$$

$$\dfrac{d\xi_n}{dt} = -\dfrac{\partial H}{\partial x_n} = 0 \,.$$

The **reduced phase space** is

$$\mathcal{U}_{red} = \{(x_1, \ldots, x_{n-1}, \xi_1, \ldots, \xi_{n-1}) \in \mathbb{R}^{2n-2} \mid \\ (x_1, \ldots, x_{n-1}, a, \xi_1, \ldots, \xi_{n-1}, c) \in \mathcal{U} \text{ for some } a\}.$$

The **reduced hamiltonian** is

$$H_{red} : \mathcal{U}_{red} \longrightarrow \mathbb{R}\,,$$
$$H_{red}(x_1, \ldots, x_{n-1}, \xi_1, \ldots, \xi_{n-1}) = H(x_1, \ldots, x_{n-1}, \xi_1, \ldots, \xi_{n-1}, c).$$

In order to find the trajectories of the original system on the hypersurface $\xi_n = c$, we look for the trajectories

$$x_1(t), \ldots, x_{n-1}(t), \xi_1(t), \ldots, \xi_{n-1}(t)$$

of the reduced system on $\mathcal{U}_{\text{red}}$. We integrate the equation

$$\frac{dx_n}{dt}(t) = \frac{\partial H}{\partial \xi_n}(x_1(t), \dots, x_{n-1}(t), \xi_1(t), \dots, \xi_{n-1}(t), c)$$

to obtain the original trajectories

$$\begin{cases} x_n(t) = x_n(0) + \int_0^t \frac{\partial H}{\partial \xi_n}(\dots)dt \\ \xi_n(t) = c. \end{cases}$$

24.3 Reduction for Product Groups

Let G_1 and G_2 be compact connected Lie groups and let $G = G_1 \times G_2$. Then

$$\mathfrak{g} = \mathfrak{g}_1 \oplus \mathfrak{g}_2 \qquad \text{and} \qquad \mathfrak{g}^* = \mathfrak{g}_1^* \oplus \mathfrak{g}_2^* .$$

Suppose that (M, ω, G, ψ) is a hamiltonian G-space with moment map

$$\psi : M \longrightarrow \mathfrak{g}_1^* \oplus \mathfrak{g}_2^* .$$

Write $\psi = (\psi_1, \psi_2)$ where $\psi_i : M \to \mathfrak{g}_i^*$ for $i = 1, 2$. The fact that ψ is equivariant implies that ψ_1 is invariant under G_2 and ψ_2 is invariant under G_1. Now reduce (M, ω) with respect to the G_1-action. Let

$$Z_1 = \psi_1^{-1}(0).$$

Assume that G_1 acts freely on Z_1. Let $M_1 = Z_1/G_1$ be the reduced space and let ω_1 be the corresponding reduced symplectic form. The action of G_2 on Z_1 commutes with the G_1-action. Since G_2 preserves ω, it follows that G_2 acts symplectically on (M_1, ω_1). Since G_1 preserves ψ_2, G_1 also preserves $\psi_2 \circ \iota_1 : Z_1 \to \mathfrak{g}_2^*$, where $\iota_1 : Z_1 \hookrightarrow M$ is inclusion. Thus $\psi_2 \circ \iota_1$ is constant on fibers of $Z_1 \overset{p_1}{\to} M_1$. We conclude that there exists a smooth map $\mu_2 : M_1 \to \mathfrak{g}_2^*$ such that $\mu_2 \circ p_1 = \psi_2 \circ \iota_1$.

Exercise. Show that:

(a) the map μ_2 is a moment map for the action of G_2 on (M_1, ω_1), and
(b) if G acts freely on $\psi^{-1}(0, 0)$, then G_2 acts freely on $\mu_2^{-1}(0)$, and there is a natural symplectomorphism

$$\mu_2^{-1}(0)/G_2 \simeq \psi^{-1}(0, 0)/G .$$

This technique of performing reduction with respect to one factor of a product group at a time is called **reduction in stages**. It may be extended to reduction by a normal subgroup $H \subset G$ and by the corresponding quotient group G/H.

24.4 Reduction at Other Levels

Suppose that a compact Lie group G acts on a symplectic manifold (M, ω) in a hamiltonian way with moment map $\mu : M \to \mathfrak{g}^*$. Let $\xi \in \mathfrak{g}^*$.

To reduce at the level ξ of μ, we need $\mu^{-1}(\xi)$ to be preserved by G, or else take the G-orbit of $\mu^{-1}(\xi)$, or else take the quotient by the maximal subgroup of G which preserves $\mu^{-1}(\xi)$.

Since μ is equivariant,

$$G \text{ preserves } \mu^{-1}(\xi) \iff G \text{ preserves } \xi$$
$$\iff \mathrm{Ad}_g^* \xi = \xi, \ \forall g \in G.$$

Of course the level 0 is always preserved. Also, when G is a torus, any level is preserved and reduction at ξ for the moment map μ, is equivalent to reduction at 0 for a shifted moment map $\phi : M \to \mathfrak{g}^*$, $\phi(p) := \mu(p) - \xi$.

Let $\mathcal{O}$ be a coadjoint orbit in $\mathfrak{g}^*$ equipped with the **canonical symplectic form** (also know as the **Kostant-Kirillov symplectic form** or the **Lie-Poisson symplectic form**) $\omega_{\mathcal{O}}$ defined in Homework 17. Let $\mathcal{O}^-$ be the orbit $\mathcal{O}$ equipped with $-\omega_{\mathcal{O}}$. The natural product action of G on $M \times \mathcal{O}^-$ is hamiltonian with moment map $\mu_{\mathcal{O}}(p, \xi) = \mu(p) - \xi$. If the Marsden-Weinstein-Meyer hypothesis is satisfied for $M \times \mathcal{O}^-$, then one obtains a **reduced space with respect to the coadjoint orbit** $\mathcal{O}$.

24.5 Orbifolds

Example. Let $G = \mathbb{T}^n$ be an n-torus. For any $\xi \in (\mathfrak{t}^n)^*$, $\mu^{-1}(\xi)$ is preserved by the $\mathbb{T}^n$-action. Suppose that ξ is a regular value of μ. (By Sard's theorem, the singular values of μ form a set of measure zero.) Then $\mu^{-1}(\xi)$ is a submanifold of codimension n. Note that

$$\xi \text{ regular} \implies d\mu_p \text{ is surjective at all } p \in \mu^{-1}(\xi)$$
$$\implies \mathfrak{g}_p = 0 \text{ for all } p \in \mu^{-1}(\xi)$$
$$\implies \text{the stabilizers on } \mu^{-1}(\xi) \text{ are finite}$$
$$\implies \mu^{-1}(\xi)/G \text{ is an \textbf{orbifold} [91, 92]}.$$

Let G_p be the stabilizer of p. By the slice theorem (Lecture 23), $\mu^{-1}(\xi)/G$ is modeled by S/G_p, where S is a G_p-invariant disk in $\mu^{-1}(\xi)$ through p and transverse to $\mathcal{O}_p$. Hence, locally $\mu^{-1}(\xi)/G$ looks indeed like $\mathbb{R}^n$ divided by a finite group action. $\diamond$

Example. Consider the S^1-action on $\mathbb{C}^2$ given by $e^{i\theta} \cdot (z_1, z_2) = (e^{ik\theta} z_1, e^{i\theta} z_2)$ for some fixed integer $k \geq 2$. This is hamiltonian with moment map

$$\mu : \quad \mathbb{C}^2 \longrightarrow \mathbb{R}$$
$$(z_1, z_2) \longmapsto -\tfrac{1}{2}(k|z_1|^2 + |z_2|^2).$$

Any $\xi < 0$ is a regular value and $\mu^{-1}(\xi)$ is a 3-dimensional ellipsoid. The stabilizer of $(z_1, z_2) \in \mu^{-1}(\xi)$ is $\{1\}$ if $z_2 \neq 0$, and is $\mathbb{Z}_k = \left\{ e^{i\frac{2\pi\ell}{k}} \mid \ell = 0, 1, \ldots, k-1 \right\}$ if $z_2 = 0$. The reduced space $\mu^{-1}(\xi)/S^1$ is called a **teardrop** orbifold or **conehead**; it has one **cone** (also known as a **dunce cap**) singularity of type k (with cone angle $\frac{2\pi}{k}$). $\diamondsuit$

Example. Let S^1 act on $\mathbb{C}^2$ by $e^{i\theta} \cdot (z_1, z_2) = (e^{ik\theta} z_1, e^{i\ell\theta} z_2)$ for some integers $k, \ell \geq 2$. Suppose that k and ℓ are relatively prime. Then

$$
\begin{aligned}
(z_1, 0) &\text{ has stabilizer } \mathbb{Z}_k \quad (\text{for } z_1 \neq 0), \\
(0, z_2) &\text{ has stabilizer } \mathbb{Z}_\ell \quad (\text{for } z_2 \neq 0), \\
(z_1, z_2) &\text{ has stabilizer } \{1\} \quad (\text{for } z_1, z_2 \neq 0).
\end{aligned}
$$

The quotient $\mu^{-1}(\xi)/S^1$ is called a **football** orbifold. It has two cone singularities, one of type k and another of type ℓ. $\diamondsuit$

Example. More generally, the reduced spaces of S^1 acting on $\mathbb{C}^n$ by

$$
e^{i\theta} \cdot (z_1, \ldots, z_n) = (e^{ik_1\theta} z_1, \ldots, e^{ik_n\theta} z_n),
$$

are called **weighted** (or **twisted**) **projective spaces**. $\diamondsuit$

Homework 18: Spherical Pendulum

This set of problems is from [53].

The **spherical pendulum** is a mechanical system consisting of a massless rigid rod of length l, fixed at one end, whereas the other end has a plumb bob of mass m, which may oscillate freely in all directions. Assume that the force of gravity is constant pointing vertically downwards, and that this is the only external force acting on this system.

Let φ, θ ($0 < \varphi < \pi$, $0 < \theta < 2\pi$) be spherical coordinates for the bob. For simplicity assume that $m = l = 1$.

1. Let η, ξ be the coordinates along the fibers of T^*S^2 induced by the spherical coordinates φ, θ on S^2. Show that the function $H : T^*S^2 \to \mathbb{R}$ given by

$$H(\varphi, \theta, \eta, \xi) = \frac{1}{2}\left(\eta^2 + \frac{\xi^2}{(\sin \varphi)^2} \right) + \cos \varphi,$$

   is an appropriate hamiltonian function to describe the spherical pendulum.
2. Compute the critical points of the function H. Show that, on S^2, there are exactly two critical points: s (where H has a minimum) and u. These points are called the **stable** and **unstable** points of H, respectively. Justify this terminology, i.e., show that a trajectory whose initial point is close to s stays close to s forever, and show that this is not the case for u. What is happening physically?
3. Show that the group of rotations about the vertical axis is a group of symmetries of the spherical pendulum.

   Show that, in the coordinates above, the integral of motion associated with these symmetries is the function

$$J(\varphi, \theta, \eta, \xi) = \xi.$$

   Give a more coordinate-independent description of J, one that makes sense also on the cotangent fibers above the North and South poles.
4. Locate all points $p \in T^*S^2$ where dH_p and dJ_p are linearly dependent:

   (a) Clearly, the two critical points s and u belong to this set. Show that these are the only two points where $dH_p = dJ_p = 0$.
   (b) Show that, if $x \in S^2$ is in the southern hemisphere ($x_3 < 0$), then there exist exactly two points, $p_+ = (x, \eta, \xi)$ and $p_- = (x, -\eta, -\xi)$, in the cotangent fiber above x where dH_p and dJ_p are linearly dependent.
   (c) Show that dH_p and dJ_p are linearly dependent along the trajectory of the hamiltonian vector field of H through p_+.

   Conclude that this trajectory is also a trajectory of the hamiltonian vector field of J, and, hence, that its projection onto S^2 is a latitudinal circle (of the form $x_3 = $ constant).

   Show that the projection of the trajectory through p_- is the same latitudinal circle traced in the opposite direction.

5. Show that any nonzero value j is a regular value of J, and that S^1 acts freely on the level set $J = j$. What happens on the cotangent fibers above the North and South poles?
6. For $j \neq 0$ describe the reduced system and sketch the level curves of the reduced hamiltonian.
7. Show that the integral curves of the original system on the level set $J = j$ can be obtained from those of the reduced system by "quadrature", in other words, by a simple integration.
8. Show that the reduced system for $j \neq 0$ has exactly one equilibrium point. Show that the corresponding relative equilibrium for the original system is one of the horizontal curves in exercise 4.
9. The **energy-momentum map** is the map $(H,J) : T^*S^2 \to \mathbb{R}^2$. Show that, if $j \neq 0$, the level set $(H,J) = (h,j)$ of the energy-momentum map is either a circle (in which case it is one of the horizontal curves in exercise 4), or a two-torus. Show that the projection onto the configuration space of the two-torus is an annular region on S^2.

Part X
Moment Maps Revisited

Moment maps and symplectic reduction have been finding infinite-dimensional incarnations with amazing consequences for differential geometry. Lecture 25 sketches the symplectic approach of Atiyah and Bott to Yang-Mills theory.

Lecture 27 describes the convexity of the image of a torus moment map, one of the most striking geometric characteristics of moment maps.

Chapter 25
Moment Map in Gauge Theory

25.1 Connections on a Principal Bundle

Let G be a Lie group and B a manifold.

Definition 25.1. A *principal G-bundle over B* is a manifold P with a smooth map $\pi : P \to B$ satisfying the following conditions:

(a) G acts freely on P (on the left),
(b) B is the orbit space for this action and π is the point-orbit projection, and
(c) there is an open covering of B, such that, to each set $\mathcal{U}$ in that covering corresponds a map $\varphi_{\mathcal{U}} : \pi^{-1}(\mathcal{U}) \to \mathcal{U} \times G$ with

$$\varphi_{\mathcal{U}}(p) = (\pi(p), s_{\mathcal{U}}(p)) \quad \text{and} \quad s_{\mathcal{U}}(g \cdot p) = g \cdot s_{\mathcal{U}}(p), \quad \forall p \in \pi^{-1}(\mathcal{U}).$$

The G-valued maps $s_{\mathcal{U}}$ are determined by the corresponding $\varphi_{\mathcal{U}}$. Condition (c) is called the property of being **locally trivial**.

If P with map $\pi : P \to B$ is a principal G-bundle over B, then the manifold B is called the **base**, the manifold P is called the **total space**, the Lie group G is called the **structure group**, and the map π is called the **projection**. This principal bundle is also represented by the following diagram:

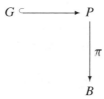

Example. Let P be the 3-sphere regarded as unit vectors in $\mathbb{C}^2$:

$$P = S^3 = \{(z_1, z_2) \in \mathbb{C}^2 : |z_1|^2 + |z_2|^2 = 1\}.$$

Let G be the circle group, where $e^{i\theta} \in S^1$ acts on S^3 by complex multiplication,

$$(z_1, z_2) \longmapsto (e^{i\theta}z_1, e^{i\theta}z_2).$$

Then the quotient space B is the first complex projective space, that is, the two-sphere. This data forms a principal S^1-bundle, known as the **Hopf fibration**:

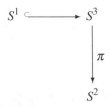

An action $\psi : G \to \mathrm{Diff}(P)$ induces an infinitesimal action

$$d\psi : \quad \mathfrak{g} \longrightarrow \chi(P)$$
$$X \longmapsto X^{\#} = \text{ vector field generated by the}$$
$$\text{one-parameter group } \{\exp tX(e) \mid t \in \mathbb{R}\}.$$

From now on, fix a basis $X_1, \ldots, X_k$ of $\mathfrak{g}$.

Let P be a principal G-bundle over B. Since the G-action is free, the vector fields $X_1^{\#}, \ldots, X_k^{\#}$ are linearly independent at each $p \in P$. The **vertical bundle** V is the rank k subbundle of TP generated by $X_1^{\#}, \ldots, X_k^{\#}$.

Exercise. Check that the vertical bundle V is the set of vectors tangent to P which lie in the kernel of the derivative of the bundle projection π. (This shows that V is independent of the choice of basis for $\mathfrak{g}$.)   ◇

Definition 25.2. A *(Ehresmann) connection* on a principal bundle P is a choice of a splitting
$$TP = V \oplus H,$$

where H is a G-invariant subbundle of TP complementary to the vertical bundle V. The bundle H is called the **horizontal bundle**.

25.2 Connection and Curvature Forms

A connection on a principal bundle P may be equivalently described in terms of 1-forms.

Definition 25.3. A *connection form* on a principal bundle P is a Lie-algebra-valued 1-form
$$A = \sum_{i=1}^{k} A_i \otimes X_i \qquad \in \Omega^1(P) \otimes \mathfrak{g}$$

such that:

(a) A is G-invariant, with respect to the product action of G on $\Omega^1(P)$ (induced by the action on P) and on $\mathfrak{g}$ (the adjoint representation), and

(b) A is vertical, in the sense that $\iota_{X^\#}A = X$ for any $X \in \mathfrak{g}$.

Exercise. Show that a connection $TP = V \oplus H$ determines a connection form A and vice-versa by the formula

$$H = \ker A = \{v \in TP \mid \iota_v A = 0\}.$$

$\diamond$

Given a connection on P, the splitting $TP = V \oplus H$ induces the following splittings for bundles:

$$T^*P = V^* \oplus H^*$$

$$\wedge^2 T^*P = (\wedge^2 V^*) \oplus (V^* \wedge H^*) \oplus (\wedge^2 H^*)$$

$$\vdots$$

and for their sections:

$$\Omega^1(P) = \Omega^1_{\text{vert}}(P) \oplus \Omega^1_{\text{horiz}}(P)$$

$$\Omega^2(P) = \Omega^2_{\text{vert}}(P) \oplus \Omega^2_{\text{mix}}(P) \oplus \Omega^2_{\text{horiz}}(P)$$

$$\vdots$$

The corresponding connection form A is in $\Omega^1_{\text{vert}} \otimes \mathfrak{g}$. Its exterior derivative dA is in

$$\Omega^2(P) \otimes \mathfrak{g} = \left(\Omega^2_{\text{vert}} \oplus \Omega^2_{\text{mix}} \oplus \Omega^2_{\text{horiz}}\right) \otimes \mathfrak{g},$$

and thus decomposes into three components,

$$dA = (dA)_{\text{vert}} + (dA)_{\text{mix}} + (dA)_{\text{horiz}} .$$

Exercise. Check that:

(a) $(dA)_{\text{vert}}(X,Y) = [X,Y]$, i.e., $(dA)_{\text{vert}} = \frac{1}{2} \sum_{i,\ell,m} c^i_{\ell m} A_\ell \wedge A_m \otimes X_i$, where the $c^i_{\ell m}$'s are the structure constants of the Lie algebra with respect to the chosen basis, and defined by $[X_\ell, X_m] = \sum_{i,\ell,m} c^i_{\ell m} X_i$;

(b) $(dA)_{\text{mix}} = 0$.

$\diamond$

According to the previous exercise, the relevance of dA may come only from its horizontal component.

Definition 25.4. The *curvature form* of a connection is the horizontal component of its connection form. I.e., if A is the connection form, then

$$\text{curv } A = (dA)_{\text{horiz}} \qquad \in \Omega^2_{\text{horiz}} \otimes \mathfrak{g}.$$

Definition 25.5. A connection is called **flat** if its curvature is zero.

25.3 Symplectic Structure on the Space of Connections

Let P be a principal G-bundle over B. If A is a connection form on P, and if $a \in \Omega^1_{\text{horiz}} \otimes \mathfrak{g}$ is G-invariant for the product action, then it is easy to check that $A + a$ is also a connection form on P. Reciprocally, any two connection forms on P differ by an $a \in (\Omega^1_{\text{horiz}} \otimes \mathfrak{g})^G$. We conclude that the set $\mathcal{A}$ of all connections on the principal G-bundle P is an affine space modeled on the linear space

$$\mathfrak{a} = (\Omega^1_{\text{horiz}} \otimes \mathfrak{g})^G .$$

Now let P be a principal G-bundle over a compact oriented 2-dimensional riemannian manifold B (for instance, B is a Riemann surface). Suppose that the group G is compact or semisimple. Atiyah and Bott [7] noticed that the corresponding space $\mathcal{A}$ of all connections may be treated as an *infinite-dimensional symplectic manifold*. This will require choosing a G-invariant inner product $\langle \cdot, \cdot \rangle$ on $\mathfrak{g}$, which always exists, either by averaging any inner product when G is compact, or by using the Killing form on semisimple groups.

Since $\mathcal{A}$ is an affine space, its tangent space at any point A is identified with the model linear space $\mathfrak{a}$. With respect to a basis $X_1, \ldots, X_k$ for the Lie algebra $\mathfrak{g}$, elements $a, b \in \mathfrak{a}^1$ are written

$$a = \sum a_i \otimes X_i \quad \text{and} \quad b = \sum b_i \otimes X_i.$$

If we wedge a and b, and then integrate over B using the riemannian volume, we obtain a real number:

$$\omega : \mathfrak{a} \times \mathfrak{a} \longrightarrow \left(\Omega^2_{\text{horiz}}(P)\right)^G \simeq \Omega^2(B) \longrightarrow \mathbb{R}$$

$$(a, b) \longmapsto \sum_{i,j} a_i \wedge b_j \langle X_i, X_j \rangle \longmapsto \int_B \sum_{i,j} a_i \wedge b_j \langle X_i, X_j \rangle.$$

We have used that the pullback $\pi^* : \Omega^2(B) \to \Omega^2(P)$ is an isomorphism onto its image $\left(\Omega^2_{\text{horiz}}(P)\right)^G$.

Exercise. Show that if $\omega(a, b) = 0$ for all $b \in \mathfrak{a}$, then a must be zero. $\diamond$

---

[1] The choice of symbols is in honor of Atiyah and Bott!

The map ω is nondegenerate, skew-symmetric, bilinear and constant in the sense that it does not depend on the base point A. Therefore, it has the right to be called a symplectic form on $\mathcal{A}$, so the pair $(\mathcal{A}, \omega)$ is an infinite-dimensional symplectic manifold.

25.4 Action of the Gauge Group

Let P be a principal G-bundle over B. A diffeomorphism $f : P \to P$ commuting with the G-action determines a diffeomorphism $f_{\text{basic}} : B \to B$ by projection.

Definition 25.6. A diffeomorphism $f : P \to P$ commuting with the G-action is a *gauge transformation* if the induced f_{basic} is the identity. The *gauge group* of P is the group $\mathcal{G}$ of all gauge transformations of P.

The derivative of an $f \in \mathcal{G}$ takes a connection $TP = V \oplus H$ to another connection $TP = V \oplus H_f$, and thus induces an action of $\mathcal{G}$ in the space $\mathcal{A}$ of all connections. Recall that $\mathcal{A}$ has a symplectic form ω. Atiyah and Bott [7] noticed that the action of $\mathcal{G}$ on $(\mathcal{A}, \omega)$ is hamiltonian, where the moment map (appropriately interpreted) is the map

$$\mu : \mathcal{A} \longrightarrow \left(\Omega^2(P) \otimes \mathfrak{g}\right)^G$$

$$A \longmapsto \operatorname{curv} A,$$

i.e., the moment map "is" the curvature! We will describe this construction in detail for the case of circle bundles in the next section.

Remark. The reduced space at level zero

$$\mathcal{M} = \mu^{-1}(0)/\mathcal{G}$$

is the space of flat connections modulo gauge equivalence, known as the **moduli space of flat connections**. It turns out that $\mathcal{M}$ is a finite-dimensional symplectic orbifold. $\diamond$

25.5 Case of Circle Bundles

What does the Atiyah-Bott construction of the previous section look like for the case when $G = S^1$?

188   25 Moment Map in Gauge Theory

Let v be the generator of the S^1-action on P, corresponding to the basis 1 of $\mathfrak{g} \simeq \mathbb{R}$. A connection form on P is a usual 1-form $A \in \Omega^1(P)$ such that

$$\mathcal{L}_v A = 0 \quad \text{and} \quad \iota_v A = 1.$$

If we fix one particular connection A_0, then any other connection is of the form $A = A_0 + a$ for some $a \in \mathfrak{a} = \left(\Omega^1_{\text{horiz}}(P)\right)^G = \Omega^1(B)$. The symplectic form on $\mathfrak{a} = \Omega^1(B)$ is simply

$$\omega : \mathfrak{a} \times \mathfrak{a} \longrightarrow \mathbb{R}$$

$$(a, b) \longmapsto \int_B \underbrace{a \wedge b}_{\in \Omega^2(B)} .$$

The gauge group is $\mathcal{G} = \text{Maps}(B, S^1)$, because a gauge transformation is multiplication by some element of S^1 over each point in B:

$$\phi : \qquad \mathcal{G} \longrightarrow \text{Diff}(P)$$

$$h : B \to S^1 \longmapsto \phi_h : P \to P$$
$$p \to h(\pi(p)) \cdot p$$

The Lie algebra of $\mathcal{G}$ is

$$\text{Lie } \mathcal{G} = \text{Maps}(B, \mathbb{R}) = C^\infty(B) .$$

Its dual space is

$$(\text{Lie } \mathcal{G})^* = \Omega^2(B) ,$$

where the duality is provided by integration over B

$$C^\infty(B) \times \Omega^2(B) \longrightarrow \mathbb{R}$$

$$(h, \beta) \longmapsto \int_B h\beta .$$

(it is topological or smooth duality, as opposed to algebraic duality) .

The gauge group acts on the space of all connections by

$$\mathcal{G} \longrightarrow \text{Diff}(\mathcal{A})$$

$$h(x) = e^{i\theta(x)} \longmapsto (A \mapsto A \underbrace{- \pi^* d\theta}_{\in \mathfrak{a}})$$

Exercise. Check the previous assertion about the action on connections.

Hint: First deal with the case where $P = S^1 \times B$ is a trivial bundle, in which case $h \in \mathcal{G}$ acts on P by

$$\phi_h : (t,x) \longmapsto (t + \theta(x), x) ,$$

and where every connection can be written $A = dt + \beta$, with $\beta \in \Omega^1(B)$. A gauge transformation $h \in \mathcal{G}$ acts on $\mathcal{A}$ by

$$A \longmapsto \phi_{h^{-1}}^*(A) .$$

$\diamond$

The infinitesimal action of $\mathcal{G}$ on $\mathcal{A}$ is

$$d\phi : \operatorname{Lie} \mathcal{G} \longrightarrow \chi(\mathcal{A})$$

$$X \longmapsto X^{\#} = \text{vector field described by the transformation}$$
$$(A \quad \mapsto \quad A \underbrace{-dX}_{\in \Omega^1(B) = \mathfrak{a}})$$

so that $X^{\#} = -dX$.

Finally, we will check that

$$\mu : \mathcal{A} \longrightarrow (\operatorname{Lie} \mathcal{G})^* = \Omega^2(B)$$

$$A \longmapsto \operatorname{curv} A$$

is indeed a moment map for the action of the gauge group on $\mathcal{A}$.

Exercise. Check that in this case:

(a) $\operatorname{curv} A = dA \quad \in \quad \left(\Omega^2_{\text{horiz}}(P)\right)^G = \Omega^2(B)$,

(b) μ is $\mathcal{G}$-invariant.

$\diamond$

The previous exercise takes care of the equivariance condition, since the action of $\mathcal{G}$ on $\Omega^2(B)$ is trivial.

Take any $X \in \operatorname{Lie} \mathcal{G} = C^\infty(B)$. We need to check that

$$d\mu^X(a) = \omega(X^{\#}, a) , \qquad \forall a \in \Omega^1(B) . \qquad (\star)$$

As for the left-hand side of $(\star)$, the map μ^X,

$$\mu^X : \mathcal{A} \longrightarrow \mathbb{R}$$
$$A \longmapsto \langle \underbrace{X}_{\in C^\infty(B)} , \underbrace{dA}_{\in \Omega^2(B)} \rangle = \int_B X \cdot dA ,$$

is linear in A. Consequently,

$$d\mu^X : \mathfrak{a} \longrightarrow \mathbb{R}$$
$$a \longmapsto \int_B X \cdot da .$$

As for the right-hand side of $(\star)$, by definition of ω, we have

$$\omega(X^{\#}, a) = \int_B X^{\#} \cdot a = -\int_B dX \cdot a \,.$$

But, by Stokes theorem, the last integral is

$$-\int_B dX \cdot a = \int_B X \cdot da \,,$$

so we are done in proving that μ is the moment map.

Homework 19: Examples of Moment Maps

1. Suppose that a Lie group G acts in a hamiltonian way on two symplectic manifolds (M_j, ω_j), $j = 1, 2$, with moment maps $\mu_j : M_j \to \mathfrak{g}^*$. Prove that the diagonal action of G on $M_1 \times M_2$ is hamiltonian with moment map $\mu : M_1 \times M_2 \to \mathfrak{g}^*$ given by

$$\mu(p_1, p_2) = \mu_1(p_1) + \mu_2(p_2), \quad \text{for } p_j \in M_j .$$

2. Let $\mathbb{T}^n = \{(t_1, \ldots, t_n) \in \mathbb{C}^n : |t_j| = 1, \text{ for all } j\}$ be a torus acting on $\mathbb{C}^n$ by

$$(t_1, \ldots, t_n) \cdot (z_1, \ldots, z_n) = (t_1^{k_1} z_1, \ldots, t_n^{k_n} z_n) ,$$

where $k_1, \ldots, k_n \in \mathbb{Z}$ are fixed. Check that this action is hamiltonian with moment map $\mu : \mathbb{C}^n \to (\mathfrak{t}^n)^* \simeq \mathbb{R}^n$ given by

$$\mu(z_1, \ldots, z_n) = -\tfrac{1}{2}(k_1|z_1|^2, \ldots, k_n|z_n|^2) \ (+ \text{ constant }).$$

3. The vector field $X^\#$ generated by $X \in \mathfrak{g}$ for the coadjoint representation of a Lie group G on $\mathfrak{g}^*$ satisfies $\langle X_\xi^\#, Y \rangle = \langle \xi, [Y, X] \rangle$, for any $Y \in \mathfrak{g}$. Equip the coadjoint orbits with the canonical symplectic forms. Show that, for each $\xi \in \mathfrak{g}^*$, the coadjoint action on the orbit $G \cdot \xi$ is hamiltonian with moment map the inclusion map:

$$\mu : G \cdot \xi \hookrightarrow \mathfrak{g}^* .$$

4. Consider the natural action of $U(n)$ on $(\mathbb{C}^n, \omega_0)$. Show that this action is hamiltonian with moment map $\mu : \mathbb{C}^n \to \mathfrak{u}(n)$ given by

$$\mu(z) = \tfrac{i}{2}zz^* ,$$

where we identify the Lie algebra $\mathfrak{u}(n)$ with its dual via the inner product $(A, B) = \text{trace}(A^*B)$.

Hint: Denote the elements of $U(n)$ in terms of real and imaginary parts $g = h + ik$. Then g acts on $\mathbb{R}^{2n}$ by the linear symplectomorphism $\begin{pmatrix} h & -k \\ k & h \end{pmatrix}$. The Lie algebra $\mathfrak{u}(n)$ is the set of skew-hermitian matrices $X = V + iW$ where $V = -V^t \in \mathbb{R}^{n \times n}$ and $W = W^t \in \mathbb{R}^{n \times n}$. Show that the infinitesimal action is generated by the hamiltonian functions

$$\mu^X(z) = -\tfrac{1}{2}(x, Wx) + (y, Vx) - \tfrac{1}{2}(y, Wy)$$

where $z = x + iy$, $x, y \in \mathbb{R}^n$ and $(\cdot, \cdot)$ is the standard inner product. Show that

$$\mu^X(z) = \tfrac{1}{2}iz^*Xz = \tfrac{1}{2}i\,\text{trace}(zz^*X) .$$

Check that μ is equivariant.

5. Consider the natural action of $U(k)$ on the space $(\mathbb{C}^{k \times n}, \omega_0)$ of complex $(k \times n)$-matrices. Identify the Lie algebra $\mathfrak{u}(k)$ with its dual via the inner product $(A, B) = \text{trace}(A^*B)$. Prove that a moment map for this action is given by

$$\mu(A) = \tfrac{i}{2}AA^* + \tfrac{\mathrm{Id}}{2i}\,, \quad \text{for } A \in \mathbb{C}^{k\times n}\,.$$

(The choice of the constant $\frac{\mathrm{Id}}{2i}$ is for convenience in Homework 20.)

Hint: Exercises 1 and 4.

6. Consider the $U(n)$-action by conjugation on the space $(\mathbb{C}^{n^2}, \omega_0)$ of complex $(n \times n)$-matrices. Show that a moment map for this action is given by

$$\mu(A) = \tfrac{i}{2}[A, A^*]\,.$$

Hint: Previous exercise and its "transpose" version.

Chapter 26
Existence and Uniqueness of Moment Maps

26.1 Lie Algebras of Vector Fields

Let (M, ω) be a symplectic manifold and $v \in \chi(M)$ a vector field on M.

$$v \text{ is symplectic} \iff \iota_v \omega \text{ is closed,}$$
$$v \text{ is hamiltonian} \iff \iota_v \omega \text{ is exact.}$$

The spaces

$$\chi^{\text{sympl}}(M) = \text{symplectic vector fields on } M,$$
$$\chi^{\text{ham}}(M) = \text{hamiltonian vector fields on } M.$$

are Lie algebras for the Lie bracket of vector fields. $C^\infty(M)$ is a Lie algebra for the Poisson bracket, $\{f, g\} = \omega(v_f, v_g)$. $H^1(M; \mathbb{R})$ and $\mathbb{R}$ are regarded as Lie algebras for the trivial bracket. We have two exact sequences of Lie algebras:

$$0 \longrightarrow \chi^{\text{ham}}(M) \hookrightarrow \chi^{\text{sympl}}(M) \longrightarrow H^1(M; \mathbb{R}) \longrightarrow 0$$
$$v \longmapsto [\iota_v \omega]$$

$$0 \longrightarrow \quad \mathbb{R} \quad \hookrightarrow \quad C^\infty(M) \longrightarrow \chi^{\text{ham}}(M) \longrightarrow 0$$
$$f \longmapsto v_f .$$

In particular, if $H^1(M; \mathbb{R}) = 0$, then $\chi^{\text{ham}}(M) = \chi^{\text{sympl}}(M)$.

Let G be a connected Lie group. A symplectic action $\psi : G \to \text{Sympl}(M, \omega)$ induces an infinitesimal action

$$d\psi : \mathfrak{g} \longrightarrow \chi^{\text{sympl}}(M)$$
$$X \longmapsto X^\# = \text{vector field generated by the}$$
$$\text{one-parameter group } \{\exp tX(e) \,|\, t \in \mathbb{R}\} .$$

Exercise. Check that the map $d\psi$ is a Lie algebra anti-homomorphism.   ◇

The action ψ is **hamiltonian** if and only if there is a Lie algebra homomorphism $\mu^* : \mathfrak{g} \to C^\infty(M)$ lifting $d\psi$, i.e., making the following diagram commute.

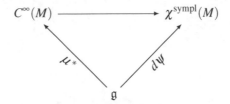

The map μ^* is then called a **comoment map** (defined in Lecture 22).

$$\text{Existence of } \mu^* \iff \text{Existence of } \mu$$
$$\text{comoment map} \qquad\qquad \text{moment map}$$

$$\text{Lie algebra homomorphism} \longleftrightarrow \text{equivariance}$$

26.2 Lie Algebra Cohomology

Let $\mathfrak{g}$ be a Lie algebra, and

$$C^k := \Lambda^k \mathfrak{g}^* = k\text{-cochains on } \mathfrak{g}$$
$$= \text{alternating } k\text{-linear maps } \underbrace{\mathfrak{g} \times \cdots \times \mathfrak{g}}_{k} \longrightarrow \mathbb{R} .$$

Define a linear operator $\delta : C^k \to C^{k+1}$ by

$$\delta c(X_0, \ldots, X_k) = \sum_{i<j} (-1)^{i+j} c([X_i, X_j], X_0, \ldots, \widehat{X_i}, \ldots, \widehat{X_j}, \ldots, X_k) .$$

Exercise. Check that $\delta^2 = 0$.   ◇

The **Lie algebra cohomology groups** (or **Chevalley cohomology groups**) of $\mathfrak{g}$ are the cohomology groups of the complex $0 \xrightarrow{\delta} C^0 \xrightarrow{\delta} C^1 \xrightarrow{\delta} \ldots$:

$$H^k(\mathfrak{g}; \mathbb{R}) := \frac{\ker \delta : C^k \longrightarrow C^{k+1}}{\operatorname{im} \delta : C^{k-1} \longrightarrow C^k} .$$

Theorem 26.1. *If $\mathfrak{g}$ is the Lie algebra of a compact connected Lie group G, then*

$$H^k(\mathfrak{g}; \mathbb{R}) = H^k_{\text{deRham}}(G) .$$

Proof. Exercise. Hint: by averaging show that the de Rham cohomology can be computed from the subcomplex of G-invariant forms.   □

Meaning of $H^1(\mathfrak{g};\mathbb{R})$ and $H^2(\mathfrak{g};\mathbb{R})$:

- An element of $C^1 = \mathfrak{g}^*$ is a linear functional on $\mathfrak{g}$. If $c \in \mathfrak{g}^*$, then $\delta c(X_0,X_1) = -c([X_0,X_1])$. The **commutator ideal** of $\mathfrak{g}$ is

$$[\mathfrak{g},\mathfrak{g}] := \{\text{linear combinations of } [X,Y] \text{ for any } X,Y \in \mathfrak{g}\} \ .$$

Since $\delta c = 0$ if and only if c vanishes on $[\mathfrak{g},\mathfrak{g}]$, we conclude that

$$H^1(\mathfrak{g};\mathbb{R}) = [\mathfrak{g},\mathfrak{g}]^0$$

where $[\mathfrak{g},\mathfrak{g}]^0 \subseteq \mathfrak{g}^*$ is the annihilator of $[\mathfrak{g},\mathfrak{g}]$.
- An element of C^2 is an alternating bilinear map $c : \mathfrak{g} \times \mathfrak{g} \to \mathbb{R}$.

$$\delta c(X_0,X_1,X_2) = -c([X_0,X_1],X_2) + c([X_0,X_2],X_1) - c([X_1,X_2],X_0) \ .$$

If $c = \delta b$ for some $b \in C^1$, then

$$c(X_0,X_1) = (\delta b)(X_0,X_1) = -b([X_0,X_1]\ ).$$

26.3 Existence of Moment Maps

Theorem 26.2. *If $H^1(\mathfrak{g};\mathbb{R}) = H^2(\mathfrak{g},\mathbb{R}) = 0$, then any symplectic G-action is hamiltonian.*

Proof. Let $\psi : G \to \mathrm{Sympl}(M,\omega)$ be a symplectic action of G on a symplectic manifold (M,ω). Since

$$H^1(\mathfrak{g};\mathbb{R}) = 0 \iff [\mathfrak{g},\mathfrak{g}] = \mathfrak{g}$$

and since commutators of symplectic vector fields are hamiltonian, we have

$$d\psi : \mathfrak{g} = [\mathfrak{g},\mathfrak{g}] \longrightarrow \chi^{\mathrm{ham}}(M).$$

The action ψ is hamiltonian if and only if there is a Lie algebra homomorphism $\mu^* : \mathfrak{g} \to C^\infty(M)$ such that the following diagram commutes.

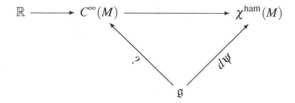

We first take an arbitrary vector space lift $\tau : \mathfrak{g} \to C^\infty(M)$ making the diagram commute, i.e., for each basis vector $X \in \mathfrak{g}$, we choose

$$\tau(X) = \tau^X \in C^\infty(M) \qquad \text{such that} \qquad v_{(\tau^X)} = d\psi(X) .$$

The map $X \mapsto \tau^X$ may not be a Lie algebra homomorphism. By construction, $\tau^{[X,Y]}$ is a hamiltonian function for $[X,Y]^\#$, and (as computed in Lecture 16) $\{\tau^X, \tau^Y\}$ is a hamiltonian function for $-[X^\#, Y^\#]$. Since $[X,Y]^\# = -[X^\#, Y^\#]$, the corresponding hamiltonian functions must differ by a constant:

$$\tau^{[X,Y]} - \{\tau^X, \tau^Y\} = c(X,Y) \in \mathbb{R} .$$

By the Jacobi identity, $\delta c = 0$. Since $H^2(\mathfrak{g}; \mathbb{R}) = 0$, there is $b \in \mathfrak{g}^*$ satisfying $c = \delta b$, $c(X,Y) = -b([X,Y])$. We define

$$\begin{aligned} \mu^* : \; &\mathfrak{g} \longrightarrow C^\infty(M) \\ &X \longmapsto \mu^*(X) = \tau^X + b(X) = \mu^X . \end{aligned}$$

Now μ^* is a Lie algebra homomorphism:

$$\mu^*([X,Y]) = \tau^{[X,Y]} + b([X,Y]) = \{\tau^X, \tau^Y\} = \{\mu^X, \mu^Y\} .$$

$\square$

So when is $H^1(\mathfrak{g}; \mathbb{R}) = H^2(\mathfrak{g}; \mathbb{R}) = 0$?

A compact Lie group G is **semisimple** if $\mathfrak{g} = [\mathfrak{g}, \mathfrak{g}]$.

Examples. The unitary group $U(n)$ is not semisimple because the multiples of the identity, $S^1 \cdot \mathrm{Id}$, form a nontrivial center; at the level of the Lie algebra, this corresponds to the 1-dimensional subspace $\mathbb{R} \cdot \mathrm{Id}$ of scalar matrices which are not commutators since they are not traceless.

Any direct product of the other compact classical groups $SU(n)$, $SO(n)$ and $Sp(n)$ is semisimple $(n > 1)$. Any commutative Lie group is not semisimple. $\Diamond$

Theorem 26.3. (Whitehead Lemmas) *Let G be a compact Lie group.*

$$G \text{ is semisimple} \qquad \Longleftrightarrow \qquad H^1(\mathfrak{g}; \mathbb{R}) = H^2(\mathfrak{g}; \mathbb{R}) = 0.$$

A proof can be found in [67, pages 93-95].

Corollary 26.4. *If G is semisimple, then any symplectic G-action is hamiltonian.*

26.4 Uniqueness of Moment Maps

Let G be a compact connected Lie group.

Theorem 26.5. *If $H^1(\mathfrak{g}; \mathbb{R}) = 0$, then moment maps for hamiltonian G-actions are unique.*

Proof. Suppose that μ_1^* and μ_2^* are two comoment maps for an action ψ:

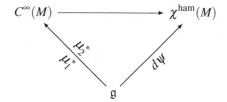

For each $X \in \mathfrak{g}$, μ_1^X and μ_2^X are both hamiltonian functions for $X^{\#}$, thus $\mu_1^X - \mu_2^X = c(X)$ is locally constant. This defines $c \in \mathfrak{g}^*$, $X \mapsto c(X)$.

Since μ_1^*, μ_2^* are Lie algebra homomorphisms, we have $c([X,Y]) = 0$, $\forall X, Y \in \mathfrak{g}$, i.e., $c \in [\mathfrak{g},\mathfrak{g}]^0 = \{0\}$. Hence, $\mu_1^* = \mu_2^*$. $\qquad\square$

Corollary of this proof. *In general, if $\mu : M \to \mathfrak{g}^*$ is a moment map, then given any $c \in [\mathfrak{g},\mathfrak{g}]^0$, $\mu_1 = \mu + c$ is another moment map.*

In other words, moment maps are unique up to elements of the dual of the Lie algebra which annihilate the commutator ideal.

The two extreme cases are:

> G semisimple: any symplectic action is hamiltonian,
> moment maps are unique.
> G commutative: symplectic actions may not be hamiltonian,
> moment maps are unique up to any constant $c \in \mathfrak{g}^*$.

Example. The circle action on $(\mathbb{T}^2, \omega = d\theta_1 \wedge d\theta_2)$ by rotations in the θ_1 direction has vector field $X^{\#} = \frac{\partial}{\partial\theta_1}$; this is a symplectic action but is not hamiltonian. $\qquad\Diamond$

Homework 20: Examples of Reduction

1. For the natural action of $U(k)$ on $\mathbb{C}^{k\times n}$ with moment map computed in exercise 5 of Homework 19, we have $\mu^{-1}(0) = \{A \in \mathbb{C}^{k\times n} \,|\, AA^* = \mathrm{Id}\}$. Show that the quotient

$$\mu^{-1}(0)/U(k) = \mathbb{G}(k,n)$$

   is the grassmannian of k-planes in $\mathbb{C}^n$.

2. Consider the S^1-action on $(\mathbb{R}^{2n+2}, \omega_0)$ which, under the usual identification of $\mathbb{R}^{2n+2}$ with $\mathbb{C}^{n+1}$, corresponds to multiplication by e^{it}. This action is hamiltonian with a moment map $\mu : \mathbb{C}^{n+1} \to \mathbb{R}$ given by

$$\mu(z) = -\tfrac{1}{2}|z|^2 + \tfrac{1}{2} \ .$$

   Prove that the reduction $\mu^{-1}(0)/S^1$ is $\mathbb{CP}^n$ with the Fubini-Study symplectic form $\omega_{\mathrm{red}} = \omega_{\mathrm{FS}}$.

   Hint: Let $\mathrm{pr} : \mathbb{C}^{n+1} \setminus \{0\} \to \mathbb{CP}^n$ denote the standard projection. Check that

$$\mathrm{pr}^* \omega_{\mathrm{FS}} = \tfrac{i}{2} \partial \bar{\partial} \log(|z|^2) \ .$$

   Prove that this form has the same restriction to S^{2n+1} as ω_0.

3. Show that the natural actions of $\mathbb{T}^{n+1}$ and $U(n+1)$ on $(\mathbb{CP}^n, \omega_{\mathrm{FS}})$ are hamiltonian, and find formulas for their moment maps.

   Hint: Previous exercise and exercises 2 and 4 of Homework 19.

Chapter 27
Convexity

27.1 Convexity Theorem

From now on, we will concentrate on actions of a torus $G = \mathbb{T}^m = \mathbb{R}^m / \mathbb{Z}^m$.

Theorem 27.1. (Atiyah [6], Guillemin-Sternberg [57]) *Let (M, ω) be a compact connected symplectic manifold, and let $\mathbb{T}^m$ be an m-torus. Suppose that $\psi : \mathbb{T}^m \to \mathrm{Sympl}(M, \omega)$ is a hamiltonian action with moment map $\mu : M \to \mathbb{R}^m$. Then:*

1. *the levels of μ are connected;*
2. *the image of μ is convex;*
3. *the image of μ is the convex hull of the images of the fixed points of the action.*

The image $\mu(M)$ of the moment map is hence called the **moment polytope**.

Proof. This proof (due to Atiyah) involves induction over $m = \dim \mathbb{T}^m$. Consider the statements:

A_m: "the levels of μ are connected, for any $\mathbb{T}^m$-action;"
B_m: "the image of μ is convex, for any $\mathbb{T}^m$-action."

Then

$$(1) \iff A_m \text{ holds for all } m,$$
$$(2) \iff B_m \text{ holds for all } m.$$

- A_1 is a non-trivial result in Morse theory.
- $A_{m-1} \implies A_m$ (induction step) is in Homework 21.
- B_1 is trivial because in $\mathbb{R}$ connectedness is convexity.
- $A_{m-1} \implies B_m$ is proved below.

Choose an injective matrix $A \in \mathbb{Z}^{m \times (m-1)}$. Consider the action of an $(m-1)$-subtorus

$$\psi_A : \mathbb{T}^{m-1} \longrightarrow \text{Sympl}(M, \omega)$$
$$\theta \longmapsto \psi_{A\theta}.$$

Exercise. The action ψ_A is hamiltonian with moment map $\mu_A = A^t \mu : M \to \mathbb{R}^{m-1}$.

$\Diamond$

Given any $p_0 \in \mu_A^{-1}(\xi)$,

$$p \in \mu_A^{-1}(\xi) \iff A^t \mu(p) = \xi = A^t \mu(p_0)$$

so that

$$\mu_A^{-1}(\xi) = \{p \in M \mid \mu(p) - \mu(p_0) \in \ker A^t\} .$$

By the first part (statement A_{m-1}), $\mu_A^{-1}(\xi)$ is connected. Therefore, if we connect p_0 to p_1 by a path p_t in $\mu_A^{-1}(\xi)$, we obtain a path $\mu(p_t) - \mu(p_0)$ in $\ker A^t$. But $\ker A^t$ is 1-dimensional. Hence, $\mu(p_t)$ must go through any convex combination of $\mu(p_0)$ and $\mu(p_1)$, which shows that any point on the line segment from $\mu(p_0)$ to $\mu(p_1)$ must be in $\mu(M)$:

$$(1-t)\mu(p_0) + t\mu(p_1) \in \mu(M), \quad 0 \leq t \leq 1.$$

Any $p_0, p_1 \in M$ can be approximated arbitrarily closely by points p_0' and p_1' with $\mu(p_1') - \mu(p_0') \in \ker A^t$ for some injective matrix $A \in \mathbb{Z}^{m \times (m-1)}$. Taking limits $p_0' \to p_0$, $p_1' \to p_1$, we obtain that $\mu(M)$ is convex.[1]

To prove part *3*, consider the fixed point set C of ψ. Homework 21 shows that C is a finite union of connected symplectic submanifolds, $C = C_1 \cup \cdots \cup C_N$. The moment map is constant on each C_j, $\mu(C_j) = \eta_j \in \mathbb{R}^m$, $j = 1, \ldots, N$. By the second part, the convex hull of $\{\eta_1, \ldots, \eta_N\}$ is contained in $\mu(M)$.

For the converse, suppose that $\xi \in \mathbb{R}^m$ and $\xi \notin$ convex hull of $\{\eta_1, \ldots, \eta_N\}$. Choose $X \in \mathbb{R}^m$ with rationally independent components and satisfying

$$\langle \xi, X \rangle > \langle \eta_j, X \rangle, \text{ for all } j .$$

By the irrationality of X, the set $\{\exp tX(e) \mid t \in \mathbb{R}\}$ is dense in $\mathbb{T}^m$, hence the zeros of the vector field $X^{\#}$ on M are the fixed points of the $\mathbb{T}^m$-action. Since $\mu^X = \langle \mu, X \rangle$ attains its maximum on one of the sets C_j, this implies

$$\langle \xi, X \rangle > \sup_{p \in M} \langle \mu(p), X \rangle ,$$

hence $\xi \notin \mu(M)$. Therefore,

$$\mu(M) = \text{ convex hull of } \{\eta_1, \ldots, \eta_N\} .$$

$\square$

---

[1] Clearly $\mu(M)$ is closed because it is compact.

27.2 Effective Actions

An action of a group G on a manifold M is called **effective** if each group element $g \neq e$ moves at least one $p \in M$, that is,

$$\bigcap_{p \in M} G_p = \{e\} \,,$$

where $G_p = \{g \in G \mid g \cdot p = p\}$ is the stabilizer of p.

Corollary 27.2. *Under the conditions of the convexity theorem, if the $\mathbb{T}^m$-action is effective, then there must be at least $m + 1$ fixed points.*

Proof. If the $\mathbb{T}^m$-action is effective, there must be a point p where the moment map is a submersion, i.e., $(d\mu_1)_p, \ldots, (d\mu_m)_p$ are linearly independent. Hence, $\mu(p)$ is an interior point of $\mu(M)$, and $\mu(M)$ is a nondegenerate convex polytope. Any nondegenerate convex polytope in $\mathbb{R}^m$ must have at least $m + 1$ vertices. The vertices of $\mu(M)$ are images of fixed points. $\square$

Theorem 27.3. *Let $(M, \omega, \mathbb{T}^m, \mu)$ be a hamiltonian $\mathbb{T}^m$-space. If the $\mathbb{T}^m$-action is effective, then $\dim M \geq 2m$.*

Proof. On an orbit $\mathcal{O}$, the moment map $\mu(\mathcal{O}) = \xi$ is constant. For $p \in \mathcal{O}$, the exterior derivative

$$d\mu_p : T_p M \longrightarrow \mathfrak{g}^*$$

maps $T_p\mathcal{O}$ to 0. Thus

$$T_p\mathcal{O} \subseteq \ker d\mu_p = (T_p\mathcal{O})^\omega \,,$$

which shows that *orbits $\mathcal{O}$ of a hamiltonian torus action are always isotropic submanifolds of M.* In particular, $\dim \mathcal{O} \leq \frac{1}{2} \dim M$.

Fact: If $\psi : \mathbb{T}^m \to \mathrm{Diff}(M)$ is an effective action, then it has orbits of dimension m; a proof may be found in [17]. $\square$

Definition 27.4. A *(symplectic) toric manifold*[2] is a compact connected symplectic manifold (M, ω) equipped with an effective hamiltonian action of a torus $\mathbb{T}$ of dimension equal to half the dimension of the manifold:

$$\dim \mathbb{T} = \frac{1}{2} \dim M$$

and with a choice of a corresponding moment map μ.

---

[2] In these notes, a toric manifold is always a *symplectic* toric manifold.

Exercise. Show that an effective hamiltonian action of a torus $\mathbb{T}^n$ on a $2n$-dimensional symplectic manifold gives rise to an integrable system.

> **Hint:** The coordinates of the moment map are commuting integrals of motion.   ◇

27.3 Examples

1. The circle $\mathbf{S}^1$ **acts on the 2-sphere** $(S^2, \omega_{\text{standard}} = d\theta \wedge dh)$ by rotations with moment map $\mu = h$ equal to the height function and moment polytope $[-1, 1]$.

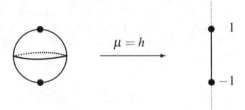

1.′ The circle $\mathbf{S}^1$ **acts on** $\mathbb{CP}^1 = \mathbb{C}^2 - 0/\sim$ with the Fubini-Study form $\omega_{\text{FS}} = \frac{1}{4}\omega_{\text{standard}}$, by $e^{i\theta} \cdot [z_0, z_1] = [z_0, e^{i\theta}z_1]$. This is hamiltonian with moment map $\mu[z_0, z_1] = -\frac{1}{2} \cdot \frac{|z_1|^2}{|z_0|^2 + |z_1|^2}$, and moment polytope $\left[-\frac{1}{2}, 0\right]$.

2. The $\mathbb{T}^2$**-action on** $\mathbb{CP}^2$ by

$$(e^{i\theta_1}, e^{i\theta_2}) \cdot [z_0, z_1, z_2] = [z_0, e^{i\theta_1}z_1, e^{i\theta_2}z_2]$$

has moment map

$$\mu[z_0, z_1, z_2] = -\frac{1}{2}\left(\frac{|z_1|^2}{|z_0|^2 + |z_1|^2 + |z_2|^2}, \frac{|z_2|^2}{|z_0|^2 + |z_1|^2 + |z_2|^2}\right) .$$

The fixed points get mapped as

$$\begin{aligned}
[1, 0, 0] &\longmapsto (0, 0) \\
[0, 1, 0] &\longmapsto \left(-\tfrac{1}{2}, 0\right) \\
[0, 0, 1] &\longmapsto \left(0, -\tfrac{1}{2}\right) .
\end{aligned}$$

Notice that the stabilizer of a preimage of the edges is S^1, while the action is free at preimages of interior points of the moment polytope.

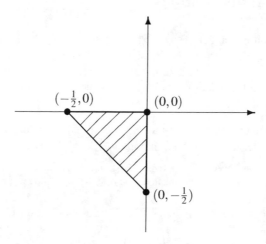

Exercise. What is the moment polytope for the $\mathbb{T}^3$-action on $\mathbb{CP}^3$ as

$$(e^{i\theta_1}, e^{i\theta_2}, e^{i\theta_3}) \cdot [z_0, z_1, z_2, z_3] = [z_0, e^{i\theta_1} z_1, e^{i\theta_2} z_2, e^{i\theta_3} z_3] \ ?$$

$\diamondsuit$

Exercise. What is the moment polytope for the $\mathbb{T}^2$-action on $\mathbb{CP}^1 \times \mathbb{CP}^1$ as

$$(e^{i\theta}, e^{i\eta}) \cdot ([z_0, z_1], [w_0, w_1]) = ([z_0, e^{i\theta} z_1], [w_0, e^{i\eta} w_1]) \ ?$$

$\diamondsuit$

Homework 21: Connectedness

Consider a hamiltonian action $\psi : \mathbb{T}^m \to \text{Sympl } (M, \omega)$, $\theta \mapsto \psi_\theta$, of an m-dimensional torus on a $2n$-dimensional compact connected symplectic manifold (M, ω). If we identify the Lie algebra of $\mathbb{T}^m$ with $\mathbb{R}^m$ by viewing $\mathbb{T}^m = \mathbb{R}^m/\mathbb{Z}^m$, and we identify the Lie algebra with its dual via the standard inner product, then the moment map for ψ is $\mu : M \to \mathbb{R}^m$.

1. Show that there exists a compatible almost complex structure J on (M, ω) which is invariant under the $\mathbb{T}^m$-action, that is, $\psi_\theta^* J = J\psi_\theta^*$, for all $\theta \in \mathbb{T}^m$.

   Hint: We cannot average almost complex structures, but we can average riemannian metrics (why?). Given a riemannian metric g_0 on M, its $\mathbb{T}^m$-average $g = \int_{\mathbb{T}^m} \psi_\theta^* g_0 d\theta$ is $\mathbb{T}^m$-invariant.

2. Show that, for any subgroup $G \subseteq \mathbb{T}^m$, the fixed-point set for G,

$$\text{Fix } (G) = \bigcap_{\theta \in G} \text{Fix } (\psi_\theta),$$

   is a symplectic submanifold of M.

   Hint: For each $p \in \text{Fix } (G)$ and each $\theta \in G$, the differential of ψ_θ at p,

$$d\psi_\theta(p) : T_pM \longrightarrow T_pM,$$

   preserves the complex structure J_p on T_pM. Consider the exponential map $\exp_p :$ $T_pM \to M$ with respect to the invariant riemannian metric $g(\cdot, \cdot) = \omega(\cdot, J\cdot)$. Show that, by uniqueness of geodesics, $\exp_p$ is equivariant, i.e.,

$$\exp_p(d\psi_\theta(p)v) = \psi_\theta(\exp_p v)$$

   for any $\theta \in G$, $v \in T_pM$. Conclude that the fixed points of ψ_θ near p correspond to the fixed points of $d\psi_\theta(p)$ on T_pM, that is

$$T_p\text{Fix } (G) = \bigcap_{\theta \in G} \ker(\text{Id} - d\psi_\theta(p)) .$$

   Since $d\psi_\theta(p) \circ J_p = J_p \circ d\psi_\theta(p)$, the eigenspace with eigenvalue 1 is invariant under J_p, and is therefore a symplectic subspace.

3. A smooth function $f : M \to \mathbb{R}$ on a compact riemannian manifold M is called a **Morse-Bott function** if its critical set $\text{Crit } (f) = \{p \in M \mid df(p) = 0\}$ is a submanifold of M and for every $p \in \text{Crit } (f)$, $T_p\text{Crit } (f) = \ker \nabla^2 f(p)$ where $\nabla^2 f(p) : T_pM \to T_pM$ denotes the linear operator obtained from the hessian via the riemannian metric. This is the natural generalization of the notion of Morse function to the case where the critical set is not just isolated points. If f is a Morse-Bott function, then $\text{Crit } (f)$ decomposes into finitely many connected critical manifolds C. The tangent space T_pM at $p \in C$ decomposes as a direct sum

$$T_pM = T_pC \oplus E_p^+ \oplus E_p^-$$

where E_p^+ and E_p^- are spanned by the positive and negative eigenspaces of $\nabla^2 f(p)$. The *index* of a connected critical submanifold C is $n_C^- = \dim E_p^-$, for any $p \in C$, whereas the *coindex* of C is $n_C^+ = \dim E_p^+$.

For each $X \in \mathbb{R}^m$, let $\mu^X = \langle \mu, X \rangle : M \to \mathbb{R}$ be the component of μ along X. Show that μ^X is a Morse-Bott function with even-dimensional critical manifolds of even index. Moreover, show that the critical set

$$\text{Crit}\,(\mu^X) = \bigcap_{\theta \in \mathbb{T}^X} \text{Fix}\,(\psi_\theta)$$

is a symplectic manifold, where $\mathbb{T}^X$ is the closure of the subgroup of $\mathbb{T}^m$ generated by X.

Hint: Assume first that X has components independent over $\mathbb{Q}$, so that $\mathbb{T}^X = \mathbb{T}^m$ and $\text{Crit}\,(\mu^X) = \text{Fix}\,(\mathbb{T}^m)$. Apply exercise 2. To prove that $T_p\text{Crit}\,(\mu^X) = \ker \nabla^2 \mu^X(p)$, show that $\ker \nabla^2 \mu^X(p) = \cap_{\theta \in \mathbb{T}^m} \ker(\text{Id} - d\psi_\theta(p))$. To see this, notice that the 1-parameter group of matrices $(d\psi_{\exp tX})_p$ coincides with $\exp(tv_p)$, where $v_p = -J_p \nabla^2 \mu^X(p) : T_pM \to T_pM$ is a vector field on T_pM. The kernel of $\nabla^2 \mu^X(p)$ corresponds to the fixed points of $d\psi_{tX}(p)$, and since X has rationally independent components, these are the common fixed points of all $d\psi_\theta(p)$, $\theta \in \mathbb{T}^m$. The eigenspaces of $\nabla^2 \mu^X(p)$ are even-dimensional because they are invariant under J_p.

4. The moment map $\mu = (\mu_1, \ldots, \mu_m)$ is called **effective** if the 1-forms $d\mu_1, \ldots, d\mu_m$ of its components are linearly independent. Show that, if μ is not effective, then the action reduces to that of an $(m-1)$-subtorus.

Hint: If μ is not effective, then the function $\mu^X = \langle \mu, X \rangle$ is constant for some nonzero $X \in \mathbb{R}^m$. Show that we can neglect the direction of X.

5. Prove that the level set $\mu^{-1}(\xi)$ is connected for every regular value $\xi \in \mathbb{R}^m$.

Hint: Prove by induction over $m = \dim \mathbb{T}^m$. For the case $m = 1$, use the lemma that all level sets $f^{-1}(c)$ of a Morse-Bott function $f : M \to \mathbb{R}$ on a compact manifold M are necessarily connected, if the critical manifolds all have index and coindex $\neq 1$ (see [83, p.178-179]). For the induction step, you can assume that ψ is effective. Then, for every $0 \neq X \in \mathbb{R}^m$, the function $\mu^X : M \to \mathbb{R}$ is not constant. Show that $\mathcal{C} := \cup_{X \neq 0}\text{Crit}\,\mu^X = \cup_{0 \neq X \in \mathbb{Z}^m}\text{Crit}\,\mu^X$ where each $\text{Crit}\,\mu^X$ is an even-dimensional proper submanifold, so the complement $M \setminus \mathcal{C}$ must be dense in M. Show that $M \setminus \mathcal{C}$ is open. Hence, by continuity, to show that $\mu^{-1}(\xi)$ is connected for every regular value $\xi = (\xi_1, \ldots, \xi_m) \in \mathbb{R}^m$, it suffices to show that $\mu^{-1}(\xi)$ is connected whenever $(\xi_1, \ldots, \xi_{m-1})$ is a regular value for a reduced moment map $(\mu_1, \ldots, \mu_{m-1})$. By the induction hypothesis, the manifold $Q = \cap_{j=1}^{m-1}\mu_j^{-1}(\xi_j)$ is connected whenever $(\xi_1, \ldots, \xi_{m-1})$ is a regular value for $(\mu_1, \ldots, \mu_{m-1})$. It suffices to show that the function $\mu_m : Q \to \mathbb{R}$ has only critical manifolds of even index and coindex (see [83, p.183]), because then, by the lemma, the level sets $\mu^{-1}(\xi) = Q \cap \mu_m^{-1}(\xi_m)$ are connected for every ξ_m.

Part XI
Symplectic Toric Manifolds

Native to algebraic geometry, toric manifolds have been studied by symplectic geometers as examples of extremely symmetric hamiltonian spaces, and as guinea pigs for new theorems. Delzant showed that symplectic toric manifolds are classified (as hamiltonian spaces) by a set of special polytopes.

Chapter 28
Classification of Symplectic Toric Manifolds

28.1 Delzant Polytopes

A $2n$-dimensional **(symplectic) toric manifold** is a compact connected symplectic manifold (M^{2n}, ω) equipped with an effective hamiltonian action of an n-torus $\mathbb{T}^n$ and with a corresponding moment map $\mu : M \to \mathbb{R}^n$.

Definition 28.1. A *Delzant polytope* Δ in $\mathbb{R}^n$ is a convex polytope satisfying:

- it is **simple**, i.e., there are n edges meeting at each vertex;
- it is **rational**, i.e., the edges meeting at the vertex p are rational in the sense that each edge is of the form $p + tu_i$, $t \geq 0$, where $u_i \in \mathbb{Z}^n$;
- it is **smooth**, i.e., for each vertex, the corresponding $u_1, \ldots, u_n$ can be chosen to be a $\mathbb{Z}$-basis of $\mathbb{Z}^n$.

Remark. The Delzant polytopes are the simple rational smooth polytopes. These are closely related to the **Newton polytopes** (which are the nonsingular n-valent polytopes), except that the vertices of a Newton polytope are required to lie on the integer lattice and for a Delzant polytope they are not. $\diamond$

Examples of Delzant polytopes:

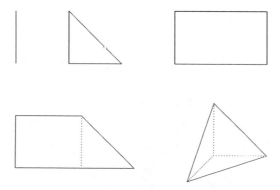

The dotted vertical line in the trapezoidal example means nothing, except that it is a picture of a rectangle plus an isosceles triangle. For "taller" triangles, smoothness would be violated. "Wider" triangles (with integral slope) may still be Delzant. The family of the Delzant trapezoids of this type, starting with the rectangle, correspond, under the Delzant construction, to **Hirzebruch surfaces**; see Homework 22.

Examples of polytopes which are **not Delzant**:

The picture on the left fails the smoothness condition, whereas the picture on the right fails the simplicity condition.

Algebraic description of Delzant polytopes:

A **facet** of a polytope is a $(n-1)$-dimensional face.

Let Δ be a Delzant polytope with $n = \dim \Delta$ and $d =$ number of facets.

A lattice vector $v \in \mathbb{Z}^n$ is **primitive** if it cannot be written as $v = ku$ with $u \in \mathbb{Z}^n$, $k \in \mathbb{Z}$ and $|k| > 1$; for instance, $(1,1)$, $(4,3)$, $(1,0)$ are primitive, but $(2,2)$, $(3,6)$ are not.

Let $v_i \in \mathbb{Z}^n$, $i = 1, \ldots, d$, be the primitive outward-pointing normal vectors to the facets.

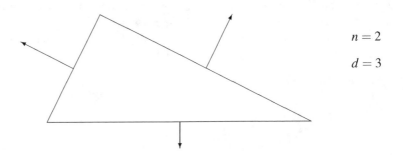

$$n = 2$$
$$d = 3$$

Then we can describe Δ as an intersection of halfspaces

$$\Delta = \{x \in (\mathbb{R}^n)^* \mid \langle x, v_i \rangle \leq \lambda_i, \ i = 1, \ldots, d\} \quad \text{for some } \lambda_i \in \mathbb{R} .$$

Example. For the picture below, we have

$$\Delta = \{x \in (\mathbb{R}^2)^* \mid x_1 \geq 0,\ x_2 \geq 0,\ x_1 + x_2 \leq 1\}$$
$$= \{x \in (\mathbb{R}^2)^* \mid \langle x, (-1,0) \rangle \leq 0,\ \langle x, (0,-1) \rangle \leq 0,\ \langle x, (1,1) \rangle \leq 1\}.$$

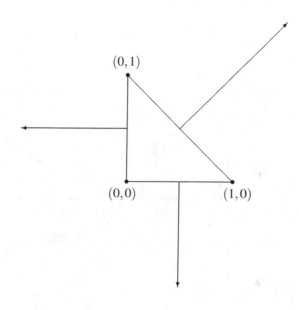

$$\Diamond$$

28.2 Delzant Theorem

We do not have a classification of symplectic manifolds, but we do have a classification of toric manifolds in terms of combinatorial data. This is the content of the Delzant theorem.

Theorem 28.2. (Delzant [23]) *Toric manifolds are classified by Delzant polytopes. More specifically, there is the following one-to-one correspondence*

$$\{toric\ manifolds\} \overset{1-1}{\longrightarrow} \{Delzant\ polytopes\}$$
$$(M^{2n}, \omega, \mathbb{T}^n, \mu) \longmapsto \mu(M).$$

We will prove the existence part (or surjectivity) in the Delzant theorem following [54]. Given a Delzant polytope, what is the corresponding toric manifold?

$$(M_\Delta, \omega_\Delta, \mathbb{T}^n, \mu) \quad \overset{?}{\longleftarrow} \quad \Delta^n$$

28.3 Sketch of Delzant Construction

Let Δ be a Delzant polytope with d facets. Let $v_i \in \mathbb{Z}^n$, $i = 1, \ldots, d$, be the primitive outward-pointing normal vectors to the facets. For some $\lambda_i \in \mathbb{R}$,

$$\Delta = \{x \in (\mathbb{R}^n)^* \mid \langle x, v_i \rangle \leq \lambda_i, \ i = 1, \ldots, d\}.$$

Let $e_1 = (1, 0, \ldots, 0), \ldots, e_d = (0, \ldots, 0, 1)$ be the standard basis of $\mathbb{R}^d$. Consider

$$\pi : \mathbb{R}^d \longrightarrow \mathbb{R}^n$$
$$e_i \longmapsto v_i .$$

Claim. The map π is onto and maps $\mathbb{Z}^d$ onto $\mathbb{Z}^n$.

Proof. The set $\{e_1, \ldots, e_d\}$ is a basis of $\mathbb{Z}^d$. The set $\{v_1, \ldots, v_d\}$ spans $\mathbb{Z}^n$ for the following reason. At a vertex p, the edge vectors $u_1, \ldots, u_n \in (\mathbb{R}^n)^*$, form a basis for $(\mathbb{Z}^n)^*$ which, without loss of generality, we may assume is the standard basis. Then the corresponding primitive normal vectors to the facets meeting at p are symmetric (in the sense of multiplication by -1) to the u_i's, hence form a basis of $\mathbb{Z}^n$. $\qquad\square$

Therefore, π induces a surjective map, still called π, between tori:

$$\begin{array}{ccc} \mathbb{R}^d/\mathbb{Z}^d & \xrightarrow{\ \pi\ } & \mathbb{R}^n/\mathbb{Z}^n \\ \| & & \| \\ \mathbb{T}^d & \longrightarrow & \mathbb{T}^n \quad \longrightarrow 0. \end{array}$$

Let
$$N = \text{kernel of } \pi \ (N \text{ is a Lie subgroup of } \mathbb{T}^d)$$
$$\mathfrak{n} = \text{Lie algebra of } N$$
$$\mathbb{R}^d = \text{Lie algebra of } \mathbb{T}^d$$
$$\mathbb{R}^n = \text{Lie algebra of } \mathbb{T}^n.$$

The exact sequence of tori

$$0 \longrightarrow N \xrightarrow{\ i\ } \mathbb{T}^d \xrightarrow{\ \pi\ } \mathbb{T}^n \longrightarrow 0$$

induces an exact sequence of Lie algebras

$$0 \longrightarrow \mathfrak{n} \xrightarrow{\ i\ } \mathbb{R}^d \xrightarrow{\ \pi\ } \mathbb{R}^n \longrightarrow 0$$

with dual exact sequence

$$0 \longrightarrow (\mathbb{R}^n)^* \xrightarrow{\ \pi^*\ } (\mathbb{R}^d)^* \xrightarrow{\ i^*\ } \mathfrak{n}^* \longrightarrow 0.$$

Now consider $\mathbb{C}^d$ with symplectic form $\omega_0 = \frac{i}{2} \sum dz_k \wedge d\bar{z}_k$, and standard hamiltonian action of $\mathbb{T}^d$

$$(e^{2\pi i t_1}, \ldots, e^{2\pi i t_d}) \cdot (z_1, \ldots, z_d) = (e^{2\pi i t_1} z_1, \ldots, e^{2\pi i t_d} z_d).$$

The moment map is $\phi : \mathbb{C}^d \longrightarrow (\mathbb{R}^d)^*$

$$\phi(z_1, \ldots, z_d) = -\pi(|z_1|^2, \ldots, |z_d|^2) + \text{constant},$$

where we choose the constant to be $(\lambda_1, \ldots, \lambda_d)$. What is the moment map for the action restricted to the subgroup N?

Exercise. Let G be any compact Lie group and H a closed subgroup of G, with $\mathfrak{g}$ and $\mathfrak{h}$ the respective Lie algebras. The inclusion $i : \mathfrak{h} \hookrightarrow \mathfrak{g}$ is dual to the projection $i^* : \mathfrak{g}^* \to \mathfrak{h}^*$. Suppose that (M, ω, G, ϕ) is a hamiltonian G-space. Show that the restriction of the G-action to H is hamiltonian with moment map

$$i^* \circ \phi : M \longrightarrow \mathfrak{h}^*.$$

$\diamondsuit$

The subtorus N acts on $\mathbb{C}^d$ in a hamiltonian way with moment map

$$i^* \circ \phi : \mathbb{C}^d \longrightarrow \mathfrak{n}^*.$$

Let $Z = (i^* \circ \phi)^{-1}(0)$ be the zero-level set.

Claim. The set Z is compact and N acts freely on Z.

This claim will be proved in the next lecture.

By the first claim, $0 \in \mathfrak{n}^*$ is a regular value of $i^* \circ \phi$. Hence, Z is a compact submanifold of $\mathbb{C}^d$ of dimension

$$\dim_{\mathbb{R}} Z = 2d - \underbrace{(d-n)}_{\dim \mathfrak{n}^*} = d + n \, .$$

The orbit space $M_\Delta = Z/N$ is a compact manifold of dimension

$$\dim_{\mathbb{R}} M_\Delta = d + n - \underbrace{(d-n)}_{\dim N} = 2n \, .$$

The point-orbit map $p : Z \to M_\Delta$ is a principal N-bundle over M_Δ. Consider the diagram

$$\begin{array}{ccc} Z & \overset{j}{\hookrightarrow} & \mathbb{C}^d \\ {\scriptstyle p} \downarrow & & \\ M_\Delta & & \end{array}$$

where $j : Z \hookrightarrow \mathbb{C}^d$ is inclusion. The Marsden-Weinstein-Meyer theorem guarantees the existence of a symplectic form ω_Δ on M_Δ satisfying

$$p^* \omega_\Delta = j^* \omega_0.$$

Exercise. Work out all details in the following simple example.

Let $\Delta = [0,a] \subset \mathbb{R}^*$ $(n=1, d=2)$. Let $v(=1)$ be the standard basis vector in $\mathbb{R}$. Then

$$\Delta : \begin{aligned} \langle x, v_1 \rangle &\leq 0 & v_1 &= -v \\ \langle x, v_2 \rangle &\leq a & v_2 &= v \,. \end{aligned}$$

The projection

$$\mathbb{R}^2 \xrightarrow{\pi} \mathbb{R}$$
$$e_1 \longmapsto -v$$
$$e_2 \longmapsto v$$

has kernel equal to the span of $(e_1 + e_2)$, so that N is the diagonal subgroup of $\mathbb{T}^2 = S^1 \times S^1$. The exact sequences become

$$0 \longrightarrow N \xrightarrow{i} \mathbb{T}^2 \xrightarrow{\pi} S^1 \longrightarrow 0$$
$$0 \longrightarrow \mathbb{R}^* \xrightarrow{\pi^*} (\mathbb{R}^2)^* \xrightarrow{i^*} \mathfrak{n}^* \longrightarrow 0$$
$$(x_1, x_2) \longmapsto x_1 + x_2 \,.$$

The action of the diagonal subgroup $N = \{(e^{2\pi i t}, e^{2\pi i t}) \in S^1 \times S^1\}$ on $\mathbb{C}^2$,

$$(e^{2\pi i t}, e^{2\pi i t}) \cdot (z_1, z_2) = (e^{2\pi i t} z_1, e^{2\pi i t} z_2) \,,$$

has moment map

$$(i^* \circ \phi)(z_1, z_2) = -\pi(|z_1|^2 + |z_2|^2) + a \,,$$

with zero-level set

$$(i^* \circ \phi)^{-1}(0) = \{(z_1, z_2) \in \mathbb{C}^2 : |z_1|^2 + |z_2|^2 = \frac{a}{\pi}\} \,.$$

Hence, the reduced space is

$$(i^* \circ \phi)^{-1}(0)/N = \mathbb{CP}^1 \qquad \text{projective space!}$$

Chapter 29
Delzant Construction

29.1 Algebraic Set-Up

Let Δ be a Delzant polytope with d facets. We can write Δ as

$$\Delta = \{x \in (\mathbb{R}^n)^* \mid \langle x, v_i \rangle \leq \lambda_i , \ i = 1, \ldots, d\},$$

for some $\lambda_i \in \mathbb{R}$. Recall the exact sequences from the previous lecture

$$0 \longrightarrow N \overset{i}{\longrightarrow} \mathbb{T}^d \overset{\pi}{\longrightarrow} \mathbb{T}^n \longrightarrow 0$$
$$0 \longrightarrow \mathfrak{n} \overset{i}{\longrightarrow} \mathbb{R}^d \overset{\pi}{\longrightarrow} \mathbb{R}^n \longrightarrow 0$$
$$e_i \longmapsto v_i$$

and the dual sequence

$$0 \longrightarrow (\mathbb{R}^n)^* \overset{\pi^*}{\longrightarrow} (\mathbb{R}^d)^* \overset{i^*}{\longrightarrow} \mathfrak{n}^* \longrightarrow 0.$$

The standard hamiltonian action of $\mathbb{T}^d$ on $\mathbb{C}^d$

$$(e^{2\pi i t_1}, \ldots, e^{2\pi i t_d}) \cdot (z_1, \ldots, z_d) = (e^{2\pi i t_1} z_1, \ldots, e^{2\pi i t_d} z_d)$$

has moment map $\phi : \mathbb{C}^d \to (\mathbb{R}^d)^*$ given by

$$\phi(z_1, \ldots, z_d) = -\pi(|z_1|^2, \ldots, |z_d|^2) + (\lambda_1, \ldots, \lambda_d).$$

The restriction of this action to N has moment map

$$i^* \circ \phi : \mathbb{C}^d \longrightarrow \mathfrak{n}^*.$$

29.2 The Zero-Level

Let $Z = (i^* \circ \phi)^{-1}(0)$.

Proposition 29.1. *The level Z is compact and N acts freely on Z.*

Proof. Let Δ' be the image of Δ by π^*. We will show that $\phi(Z) = \Delta'$. Since ϕ is a proper map and Δ' is compact, it will follow that Z is compact.

Lemma 29.2. *Let $y \in (\mathbb{R}^d)^*$. Then:*

$$y \in \Delta' \iff y \text{ is in the image of } Z \text{ by } \phi.$$

Proof of the lemma. The value y is in the image of Z by ϕ if and only if both of the following conditions hold:

1. y is in the image of ϕ;
2. $i^* y = 0$.

Using the expression for ϕ and the third exact sequence, we see that these conditions are equivalent to:

1. $\langle y, e_i \rangle \leq \lambda_i$ for $i = 1, \ldots, d$.
2. $y = \pi^*(x)$ for some $x \in (\mathbb{R}^n)^*$.

Suppose that the second condition holds, so that $y = \pi^*(x)$. Then

$$\langle y, e_i \rangle \leq \lambda_i, \forall i \iff \langle \pi^*(x), e_i \rangle \leq \lambda_i, \forall i$$
$$\iff \langle x, \pi(e_i) \rangle \leq \lambda_i, \forall i$$
$$\iff x \in \Delta.$$

Thus, $y \in \phi(Z) \iff y \in \pi^*(\Delta) = \Delta'$.   □

Hence, we have a surjective proper map $\phi : Z \to \Delta'$. Since Δ' is compact, we conclude that Z is compact. It remains to show that N acts freely on Z.

We define a stratification of Z with three equivalent descriptions:

- Define a stratification on Δ' whose ith stratum is the closure of the union of the i-dimensional faces of Δ'. Pull this stratification back to Z by ϕ.
  We can obtain a more explicit description of the stratification on Z:
- Let F be a face of Δ' with $\dim F = n - r$. Then F is characterized (as a subset of Δ') by r equations

$$\langle y, e_i \rangle = \lambda_i , \quad i = i_1, \ldots, i_r .$$

We write $F = F_I$ where $I = (i_1, \ldots, i_r)$ has $1 \leq i_1 < i_2 \ldots < i_r \leq d$.
Let $z = (z_1, \ldots, z_d) \in Z$.

$$z \in \phi^{-1}(F_I) \iff \phi(z) \in F_I$$
$$\iff \langle \phi(z), e_i \rangle = \lambda_i, \quad \forall i \in I$$
$$\iff -\pi |z_i|^2 + \lambda_i = \lambda_i, \quad \forall i \in I$$
$$\iff z_i = 0, \quad \forall i \in I.$$

- The $\mathbb{T}^d$-action on $\mathbb{C}^d$ preserves ϕ, so the $\mathbb{T}^d$-action takes $Z = \phi^{-1}(\Delta')$ onto itself, so $\mathbb{T}^d$ acts on Z.

Exercise. The stratification of Z is just the stratification of Z into $\mathbb{T}^d$ orbit types. More specifically, if $z \in Z$ and $\phi(z) \in F_I$ then the stabilizer of z in $\mathbb{T}^d$ is $(\mathbb{T}^d)_I$ where

$$I = (i_1, \ldots, i_r),$$
$$F_I = \{ y \in \Delta' \mid \langle y, e_i \rangle = \lambda_i, \forall i \in I \},$$

and

$$(\mathbb{T}^d)_I = \{ (e^{2\pi i t_1}, \ldots, e^{2\pi i t_d}) \mid e^{2\pi i t_s} = 1, \forall s \notin I \}$$

Hint: Suppose that $z = (z_1, \ldots, z_d) \in \mathbb{C}^d$. Then

$$(e^{2\pi i t_1} z_1, \ldots, e^{2\pi i t_d} z_d) = (z_1, \ldots, z_d)$$

if and only if $e^{2\pi i t_s} = 1$ whenever $z_s \neq 0$.

$$\diamondsuit$$

In order to show that N acts freely on Z, consider the worst case scenario of points $z \in Z$ whose stabilizer under the action of $\mathbb{T}^d$ is a large as possible. Now $(\mathbb{T}^d)_I$ is largest when $F_I = \{y\}$ is a vertex of Δ'. Then y satisfies n equations

$$\langle y, e_i \rangle = \lambda_i, \quad i \in I = \{i_1, \ldots, i_n\}.$$

Lemma 29.3. *Let* $z \in Z$ *be such that* $\phi(z)$ *is a vertex of* Δ'. *Let* $(\mathbb{T}^d)_I$ *be the stabilizer of* z. *Then the map* $\pi : \mathbb{T}^d \to \mathbb{T}^n$ *maps* $(\mathbb{T}^d)_I$ *bijectively onto* $\mathbb{T}^n$.

Since $N = \ker \pi$, this lemma shows that in the worst case, the stabilizer of z intersects N in the trivial group. It will follow that N acts freely at this point and hence on Z.

Proof of the lemma. Suppose that $\phi(z) = y$ is a vertex of Δ'. Renumber the indices so that

$$I = (1, 2, \ldots, n).$$

Then

$$(\mathbb{T}^d)_I = \{ (e^{2\pi i t_1}, \ldots, e^{2\pi i t_n}, 1, \ldots, 1) \mid t_i \in \mathbb{R} \}.$$

The hyperplanes meeting at y are

$$\langle y', e_i \rangle = \lambda_i, \quad i = 1, \ldots, n.$$

By definition of Delzant polytope, the set $\pi(e_1), \ldots, \pi(e_n)$ is a basis of $\mathbb{Z}^n$. Thus, $\pi : (\mathbb{T}^d)_I \to \mathbb{T}^n$ is bijective.   □

This proves the theorem in the worst case scenario, and hence in general.   □

29.3 Conclusion of the Delzant Construction

We continue the construction of $(M_\Delta, \omega_\Delta)$ from Δ. We already have that

$$M_\Delta = Z/N$$

is a compact $2n$-dimensional manifold. Let ω_Δ be the reduced symplectic form.

Claim. The manifold $(M_\Delta, \omega_\Delta)$ is a hamiltonian $\mathbb{T}^n$-space with a moment map μ having image $\mu(M_\Delta) = \Delta$.

Suppose that $z \in Z$. The stabilizer of z with respect to the $\mathbb{T}^d$-action is $(\mathbb{T}^d)_I$, and

$$(\mathbb{T}^d)_I \cap N = \{e\} \ .$$

In the worst case scenario, F_I is a vertex of Δ' and $(\mathbb{T}^d)_I$ is an n-dimensional subgroup of $\mathbb{T}^d$. In any case, there is a right inverse map $\pi^{-1} : \mathbb{T}^n \to (\mathbb{T}^d)_I$. Thus, the exact sequence

$$0 \longrightarrow N \longrightarrow \mathbb{T}^d \longrightarrow \mathbb{T}^n \longrightarrow 0$$

splits, and $\mathbb{T}^d = N \times \mathbb{T}^n$.

Apply the results on reduction for product groups (Section 24.3) to our situation of $\mathbb{T}^d = N \times \mathbb{T}^n$ acting on $(M_\Delta, \omega_\Delta)$. The moment map is

$$\phi : \mathbb{C}^d \longrightarrow (\mathbb{R}^d)^* = \mathfrak{n}^* \oplus (\mathbb{R}^n)^* \ .$$

Let $j : Z \hookrightarrow \mathbb{C}^d$ be the inclusion map, and let

$$\mathrm{pr}_1 : (\mathbb{R}^d)^* \longrightarrow \mathfrak{n}^* \qquad \text{and} \qquad \mathrm{pr}_2 : (\mathbb{R}^d)^* \longrightarrow (\mathbb{R}^n)^*$$

be the projection maps. The map

$$\mathrm{pr}_2 \circ \phi \circ j : Z \longrightarrow (\mathbb{R}^n)^*$$

is constant on N-orbits. Thus there exists a map

$$\mu : M_\Delta \longrightarrow (\mathbb{R}^n)^*$$

such that

$$\mu \circ p = \mathrm{pr}_2 \circ \phi \circ j \ .$$

The image of μ is equal to the image of $\mathrm{pr}_2 \circ \phi \circ j$. We showed earlier that $\phi(Z) = \Delta'$. Thus

$$\text{Image of } \mu = \mathrm{pr}_2(\Delta') = \underbrace{\mathrm{pr}_2 \circ \pi^*}_{\text{id}}(\Delta) = \Delta.$$

Thus $(M_\Delta, \omega_\Delta)$ is the required toric manifold corresponding to Δ.

29.4 Idea Behind the Delzant Construction

We use the idea that $\mathbb{R}^d$ is "universal" in the sense that any n-dimensional polytope Δ with d facets can be obtained by intersecting the negative orthant $\mathbb{R}^d_-$ with an affine plane A. Given Δ, to construct A first write Δ as:

$$\Delta = \{x \in \mathbb{R}^n \mid \langle x, v_i \rangle \leq \lambda_i, \ i = 1, \ldots, d\}.$$

Define

$$\pi : \mathbb{R}^d \longrightarrow \mathbb{R}^n \quad \text{with dual map} \quad \pi^* : \mathbb{R}^n \longrightarrow \mathbb{R}^d.$$
$$e_i \longmapsto v_i$$

Then

$$\pi^* - \lambda : \mathbb{R}^n \longrightarrow \mathbb{R}^d$$

is an affine map, where $\lambda = (\lambda_1, \ldots, \lambda_d)$. Let A be the image of $\pi^* - \lambda$. Then A is an n-dimensional affine plane.

Claim. We have the equality $(\pi^* - \lambda)(\Delta) = \mathbb{R}^d_- \cap A$.

Proof. Let $x \in \mathbb{R}^n$. Then

$$\begin{aligned}
(\pi^* - \lambda)(x) \in \mathbb{R}^d_- &\iff \langle \pi^*(x) - \lambda, e_i \rangle \leq 0, \forall i \\
&\iff \langle x, \pi(e_i) \rangle - \lambda_i \leq 0, \forall i \\
&\iff \langle x, v_i \rangle \leq \lambda_i, \forall i \\
&\iff x \in \Delta.
\end{aligned}$$

$\square$

We conclude that $\Delta \simeq \mathbb{R}^d_- \cap A$. Now $\mathbb{R}^d_-$ is the image of the moment map for the standard hamiltonian action of $\mathbb{T}^d$ on $\mathbb{C}^d$

$$\phi : \mathbb{C}^d \longrightarrow \mathbb{R}^d$$
$$(z_1, \ldots, z_d) \longmapsto -\pi(|z_1|^2, \ldots, |z_d|^2).$$

Facts.

- The set $\phi^{-1}(A) \subset \mathbb{C}^d$ is a compact submanifold. Let $i : \phi^{-1}(A) \hookrightarrow \mathbb{C}^d$ denote inclusion. Then $i^* \omega_0$ is a closed 2-form which is degenerate. Its kernel is an integrable distribution. The corresponding foliation is called the **null foliation**.

- The null foliation of $i^* \omega_0$ is a principal fibration, so we take the quotient:

$$N \hookrightarrow \phi^{-1}(A)$$
$$\downarrow$$
$$M_\Delta \quad = \phi^{-1}(A)/N$$

Let ω_Δ be the reduced symplectic form.
- The (non-effective) action of $\mathbb{T}^d = N \times \mathbb{T}^n$ on $\phi^{-1}(A)$ has a "moment map" with image $\phi(\phi^{-1}(A)) = \Delta$. (By "moment map" we mean a map satisfying the usual definition even though the closed 2-form is not symplectic.)

Theorem 29.4. *For any $x \in \Delta$, we have that $\mu^{-1}(x)$ is a single $\mathbb{T}^n$-orbit.*

Proof. Exercise.
First consider the standard $\mathbb{T}^d$-action on $\mathbb{C}^d$ with moment map $\phi : \mathbb{C}^d \to \mathbb{R}^d$. Show that $\phi^{-1}(y)$ is a single $\mathbb{T}^d$-orbit for any $y \in \phi(\mathbb{C}^d)$. Now observe that

$$y \in \Delta' = \pi^*(\Delta) \iff \phi^{-1}(y) \subseteq Z.$$

Suppose that $y = \pi^*(x)$. Show that $\mu^{-1}(x) = \phi^{-1}(y)/N$. But $\phi^{-1}(y)$ is a single $\mathbb{T}^d$-orbit where $\mathbb{T}^d = N \times \mathbb{T}^n$, hence $\mu^{-1}(x)$ is a single $\mathbb{T}^n$-orbit. $\quad\square$

Therefore, for toric manifolds, Δ is the orbit space.
Now Δ is a *manifold with corners*. At every point p in a face F, the tangent space $T_p \Delta$ is the subspace of $\mathbb{R}^n$ tangent to F. We can visualize $(M_\Delta, \omega_\Delta, \mathbb{T}^n, \mu)$ from Δ as follows. First take the product $\mathbb{T}^n \times \Delta$. Let p lie in the interior of $\mathbb{T}^n \times \Delta$. The tangent space at p is $\mathbb{R}^n \times (\mathbb{R}^n)^*$. Define ω_p by:

$$\omega_p(v, \xi) = \xi(v) = -\omega_p(\xi, v) \quad \text{and} \quad \omega_p(v, v') = \omega(\xi, \xi') = 0.$$

for all $v, v' \in \mathbb{R}^n$ and $\xi, \xi' \in (\mathbb{R}^n)^*$. Then ω is a closed nondegenerate 2-form on the interior of $\mathbb{T}^n \times \Delta$. At the corner there are directions missing in $(\mathbb{R}^n)^*$, so ω is a degenerate pairing. Hence, we need to eliminate the corresponding directions in $\mathbb{R}^n$. To do this, we collapse the orbits corresponding to subgroups of $\mathbb{T}^n$ generated by directions orthogonal to the annihilator of that face.

Example. Consider
$$(S^2, \omega = d\theta \wedge dh, S^1, \mu = h),$$

where S^1 acts on S^2 by rotation. The image of μ is the line segment $I = [-1, 1]$. The product $S^1 \times I$ is an open-ended cylinder. By collapsing each end of the cylinder to a point, we recover the 2-sphere. $\quad\diamond$

Exercise. Build $\mathbb{CP}^2$ from $\mathbb{T}^2 \times \Delta$ where Δ is a right-angled isosceles triangle. $\quad\diamond$

Finally, $\mathbb{T}^n$ acts on $\mathbb{T}^n \times \Delta$ by multiplication on the $\mathbb{T}^n$ factor. The moment map for this action is projection onto the Δ factor.

Homework 22: Delzant Theorem

1.(a) Consider the standard $(S^1)^3$-action on $\mathbb{CP}^3$:

$$(e^{i\theta_1}, e^{i\theta_2}, e^{i\theta_3}) \cdot [z_0, z_1, z_2, z_3] = [z_0, e^{i\theta_1} z_1, e^{i\theta_2} z_2, e^{i\theta_3} z_3] \ .$$

Exhibit explicitly the subsets of $\mathbb{CP}^3$ for which the stabilizer under this action is $\{1\}$, S^1, $(S^1)^2$ and $(S^1)^3$. Show that the images of these subsets under the moment map are the interior, the facets, the edges and the vertices, respectively.

(b) Classify all 2-dimensional Delzant polytopes with 4 vertices, up to translation and the action of $SL(2;\mathbb{Z})$.

Hint: By a linear transformation in $SL(2;\mathbb{Z})$, you can make one of the angles in the polytope into a square angle. Check that automatically another angle also becomes 90°.

(c) What are all the 4-dimensional symplectic toric manifolds that have four fixed points?

2. Take a Delzant polytope in $\mathbb{R}^n$ with a vertex p and with primitive (inward-pointing) edge vectors $u_1, \ldots, u_n$ at p. Chop off the corner to obtain a new polytope with the same vertices except p, and with p replaced by n new vertices:

$$p + \varepsilon u_j , \quad j = 1, \ldots, n ,$$

where ε is a small positive real number. Show that this new polytope is also Delzant. The corresponding toric manifold is the ε-**symplectic blowup** of the original one.

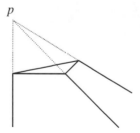

3. The toric 4-manifold $\mathcal{H}_n$ corresponding to the polygon with vertices $(0,0)$, $(n+1,0)$, $(0,1)$ and $(1,1)$, for n a nonnegative integer, is called a **Hirzebruch surface**.

(a) What is the manifold $\mathcal{H}_0$? What is the manifold $\mathcal{H}_1$?

    Hint:

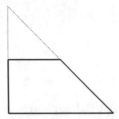

(b) Construct the manifold $\mathcal{H}_n$ by symplectic reduction of $\mathbb{C}^4$ with respect to an action of $(S^1)^2$.

(c) Exhibit $\mathcal{H}_n$ as a $\mathbb{CP}^1$-bundle over $\mathbb{CP}^1$.

4. Which $2n$-dimensional toric manifolds have exactly $n+1$ fixed points?

Chapter 30
Duistermaat-Heckman Theorems

30.1 Duistermaat-Heckman Polynomial

Let (M^{2n}, ω) be a symplectic manifold. Then $\frac{\omega^n}{n!}$ is the symplectic volume form.

Definition 30.1. The *Liouville measure* (or *symplectic measure*) of a Borel subset[1] $\mathcal{U}$ of M is

$$m_\omega(\mathcal{U}) = \int_\mathcal{U} \frac{\omega^n}{n!} .$$

Let G be a torus. Suppose that (M, ω, G, μ) is a hamiltonian G-space, and that the moment map μ is proper.

Definition 30.2. The *Duistermaat-Heckman measure*, m_{DH}, on $\mathfrak{g}^*$ is the push-forward of m_ω by $\mu : M \to \mathfrak{g}^*$. That is,

$$m_{DH}(U) = (\mu_* m_\omega)(U) = \int_{\mu^{-1}(U)} \frac{\omega^n}{n!}$$

for any Borel subset U of $\mathfrak{g}^*$.

For a compactly-supported function $h \in C^\infty(\mathfrak{g}^*)$, we define its integral with respect to the Duistermaat-Heckman measure to be

$$\int_{\mathfrak{g}^*} h \, dm_{DH} = \int_M (h \circ \mu) \frac{\omega^n}{n!} .$$

On $\mathfrak{g}^*$ regarded as a vector space, say $\mathbb{R}^n$, there is also the Lebesgue (or euclidean) measure, m_0. The relation between m_{DH} and m_0 is governed by the Radon-Nikodym derivative, denoted by $\frac{dm_{DH}}{dm_0}$, which is a generalized function satisfying

$$\int_{\mathfrak{g}^*} h \, dm_{DH} = \int_{\mathfrak{g}^*} h \frac{dm_{DH}}{dm_0} \, dm_0 .$$

---

[1] The set $\mathcal{B}$ of **Borel subsets** is the σ-*ring* generated by the set of compact subsets, i.e., if $A, B \in \mathcal{B}$, then $A \setminus B \in \mathcal{B}$, and if $A_i \in \mathcal{B}$, $i = 1, 2, \ldots$, then $\cup_{i=1}^\infty A_i \in \mathcal{B}$.

Theorem 30.3. (Duistermaat-Heckman, 1982 [31])   *The Duistermaat-Heckman measure is a piecewise polynomial multiple of Lebesgue (or euclidean) measure m_0 on $\mathfrak{g}^* \simeq \mathbb{R}^n$, that is, the Radon-Nikodym derivative*

$$f = \frac{dm_{DH}}{dm_0}$$

is piecewise polynomial. More specifically, for any Borel subset U of $\mathfrak{g}^*$,

$$m_{DH}(U) = \int_U f(x)\,dx\ ,$$

where $dx = dm_0$ is the Lebesgue volume form on U and $f : \mathfrak{g}^* \simeq \mathbb{R}^n \to \mathbb{R}$ is polynomial on any region consisting of regular values of μ.

The proof of Theorem 30.3 for the case $G = S^1$ is in Section 30.3. The proof for the general case, which follows along similar lines, can be found in, for instance, [54], besides the original articles.

The Radon-Nikodym derivative f is called the **Duistermaat-Heckman polynomial**. In the case of a toric manifold, the Duistermaat-Heckman polynomial is a universal constant equal to $(2\pi)^n$ when Δ is n-dimensional. Thus the symplectic volume of $(M_\Delta, \omega_\Delta)$ is $(2\pi)^n$ times the euclidean volume of Δ.

Example. Consider $(S^2, \omega = d\theta \wedge dh, S^1, \mu = h)$. The image of μ is the interval $[-1,1]$. The Lebesgue measure of $[a,b] \subseteq [-1,1]$ is

$$m_0([a,b]) = b - a\ .$$

The Duistermaat-Heckman measure of $[a,b]$ is

$$m_{DH}([a,b]) = \int_{\{(\theta,h)\in S^2 \mid a \leq h \leq b\}} d\theta\,dh = 2\pi(b - a)\ .$$

Consequently, the spherical area between two horizontal circles depends only on the vertical distance between them, a result which was known to Archimedes around 230 BC.

Corollary 30.4. *For the standard hamiltonian action of S^1 on (S^2, ω), we have*

$$m_{DH} = 2\pi\, m_0\ .$$

30.2 Local Form for Reduced Spaces

Let (M, ω, G, μ) be a hamiltonian G-space, where G is an n-torus.[2] Assume that μ is proper. If G acts freely on $\mu^{-1}(0)$, it also acts freely on nearby levels $\mu^{-1}(t)$, $t \in \mathfrak{g}^*$ and $t \approx 0$. Consider the reduced spaces

$$M_{\text{red}} = \mu^{-1}(0)/G \qquad \text{and} \qquad M_t = \mu^{-1}(t)/G$$

with reduced symplectic forms ω_{red} and ω_t. What is the relation between these reduced spaces as symplectic manifolds?

For simplicity, we will assume G to be the circle S^1. Let $Z = \mu^{-1}(0)$ and let $i : Z \hookrightarrow M$ be the inclusion map. We fix a connection form $\alpha \in \Omega^1(Z)$ for the principal bundle

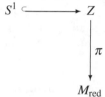

that is, $\mathcal{L}_{X^\#}\alpha = 0$ and $\iota_{X^\#}\alpha = 1$, where $X^\#$ is the infinitesimal generator for the S^1-action. From α we construct a 2-form on the product manifold $Z \times (-\varepsilon, \varepsilon)$ by the recipe

$$\sigma = \pi^*\omega_{\text{red}} - d(x\alpha) ,$$

x being a linear coordinate on the interval $(-\varepsilon, \varepsilon) \subset \mathbb{R} \simeq \mathfrak{g}^*$. (By abuse of notation, we shorten the symbols for forms on $Z \times (-\varepsilon, \varepsilon)$ which arise by pullback via projection onto each factor.)

Lemma 30.5. *The 2-form σ is symplectic for ε small enough.*

Proof. The form σ is clearly closed. At points where $x = 0$, we have

$$\sigma|_{x=0} = \pi^*\omega_{\text{red}} + \alpha \wedge dx ,$$

which satisfies

$$\sigma|_{x=0}\left(X^\#, \frac{\partial}{\partial x}\right) = 1 ,$$

so σ is nondegenerate along $Z \times \{0\}$. Since nondegeneracy is an open condition, we conclude that σ is nondegenerate for x in a sufficiently small neighborhood of 0. $\square$

Notice that σ is invariant with respect to the S^1-action on the first factor of $Z \times (-\varepsilon, \varepsilon)$. In fact, this S^1-action is hamiltonian with moment map given by projection onto the second factor,

---

[2] The discussion in this section may be extended to hamiltonian actions of other compact Lie groups, not necessarily tori; see [54, Exercises 2.1-2.10].

$$x : Z \times (-\varepsilon, \varepsilon) \longrightarrow (-\varepsilon, \varepsilon) ,$$

as is easily verified:

$$\iota_{X^\#} \sigma = -\iota_{X^\#} d(x\alpha) = -\underbrace{\mathcal{L}_{X^\#}(x\alpha)}_{0} + d\underbrace{\iota_{X^\#}(x\alpha)}_{x} = dx .$$

Lemma 30.6. *There exists an equivariant symplectomorphism between a neighborhood of Z in M and a neighborhood of $Z \times \{0\}$ in $Z \times (-\varepsilon, \varepsilon)$, intertwining the two moment maps, for ε small enough.*

Proof. The inclusion $i_0 : Z \hookrightarrow Z \times (-\varepsilon, \varepsilon)$ as $Z \times \{0\}$ and the natural inclusion $i : Z \hookrightarrow M$ are S^1-equivariant coisotropic embeddings. Moreover, they satisfy $i_0^* \sigma = i^* \omega$ since both sides are equal to $\pi^* \omega_{\text{red}}$, and the moment maps coincide on Z because $i_0^* x = 0 = i^* \mu$. Replacing ε by a smaller positive number if necessary, the result follows from the equivariant version of the coisotropic embedding theorem stated in Section 8.3.[3]   □

Therefore, in order to compare the reduced spaces

$$M_t = \mu^{-1}(t)/S^1 , \qquad t \approx 0 ,$$

we can work in $Z \times (-\varepsilon, \varepsilon)$ and compare instead the reduced spaces

$$x^{-1}(t)/S^1 , \qquad t \approx 0 .$$

Proposition 30.7. *The reduced space (M_t, ω_t) is symplectomorphic to*

$$(M_{\text{red}}, \omega_{\text{red}} - t\beta) ,$$

where β is the curvature form of the connection α.

Proof. By Lemma 30.6, (M_t, ω_t) is symplectomorphic to the reduced space at level t for the hamiltonian space $(Z \times (-\varepsilon, \varepsilon), \sigma, S^1, x)$. Since $x^{-1}(t) = Z \times \{t\}$, where S^1 acts on the first factor, all the manifolds $x^{-1}(t)/S^1$ are diffeomorphic to $Z/S^1 = M_{\text{red}}$. As for the symplectic forms, let $\iota_t : Z \times \{t\} \hookrightarrow Z \times (-\varepsilon, \varepsilon)$ be the inclusion map. The restriction of σ to $Z \times \{t\}$ is

$$\iota_t^* \sigma = \pi^* \omega_{\text{red}} - t d\alpha .$$

---

[3] The equivariant version of Theorem 8.6 needed for this purpose may be phrased as follows: *Let (M_0, ω_0), (M_1, ω_1) be symplectic manifolds of dimension $2n$, G a compact Lie group acting on (M_i, ω_i), $i = 0, 1$, in a hamiltonian way with moment maps μ_0 and μ_1, respectively, Z a manifold of dimension $k \geq n$ with a G-action, and $\iota_i : Z \hookrightarrow M_i$, $i = 0, 1$, G-equivariant coisotropic embeddings. Suppose that $\iota_0^* \omega_0 = \iota_1^* \omega_1$ and $\iota_0^* \mu_0 = \iota_1^* \mu_1$. Then there exist G-invariant neighborhoods $\mathcal{U}_0$ and $\mathcal{U}_1$ of $\iota_0(Z)$ and $\iota_1(Z)$ in M_0 and M_1, respectively, and a G-equivariant symplectomorphism $\varphi : \mathcal{U}_0 \to \mathcal{U}_1$ such that $\varphi \circ \iota_0 = \iota_1$ and $\mu_0 = \varphi^* \mu_1$.*

By definition of curvature, $d\alpha = \pi^*\beta$. Hence, the reduced symplectic form on $x^{-1}(t)/S^1$ is

$$\omega_{\text{red}} - t\beta \ .$$

$\square$

In loose terms, Proposition 30.7 says that the reduced forms ω_t vary linearly in t, for t close enough to 0. However, the identification of M_t with M_{red} as abstract manifolds is not natural. Nonetheless, any two such identifications are isotopic. By the homotopy invariance of de Rham classes, we obtain:

Theorem 30.8. (Duistermaat-Heckman, 1982 [31]) *The cohomology class of the reduced symplectic form* $[\omega_t]$ *varies linearly in t. More specifically,*

$$[\omega_t] = [\omega_{\text{red}}] + tc \ ,$$

where $c = [-\beta] \in H^2_{\text{deRham}}(M_{\text{red}})$ *is the first Chern class of the* S^1-*bundle* $Z \to M_{\text{red}}$.

Remark on conventions. Connections on principal bundles are Lie algebra-valued 1-forms; cf. Section 25.2. Often the Lie algebra of S^1 is identified with $2\pi i\mathbb{R}$ under the exponential map $\exp : \mathfrak{g} \simeq 2\pi i\mathbb{R} \to S^1$, $\xi \mapsto e^\xi$. Given a principal S^1-bundle, by this identification the infinitesimal action maps the generator $2\pi i$ of $2\pi i\mathbb{R}$ to the generating vector field $X^\#$. A connection form A is then an imaginary-valued 1-form on the total space satisfying $\mathcal{L}_{X^\#}A = 0$ and $\iota_{X^\#}A = 2\pi i$. Its curvature form B is an imaginary-valued 2-form on the base satisfying $\pi^*B = dA$. By the Chern-Weil isomorphism, the **first Chern class** of the principal S^1-bundle is $c = [\frac{i}{2\pi}B]$.

In this lecture, we identify the Lie algebra of S^1 with $\mathbb{R}$ and implicitly use the exponential map $\exp : \mathfrak{g} \sim \mathbb{R} \to S^1$, $t \mapsto e^{2\pi it}$. Hence, given a principal S^1-bundle, the infinitesimal action maps the generator 1 of $\mathbb{R}$ to $X^\#$, and here a connection form α is an ordinary 1-form on the total space satisfying $\mathcal{L}_{X^\#}\alpha = 0$ and $\iota_{X^\#}\alpha = 1$. The curvature form β is an ordinary 2-form on the base satisfying $\pi^*\beta = d\alpha$. Consequently, we have $A = 2\pi i\alpha$, $B = 2\pi i\beta$ and the first Chern class is given by $c = [-\beta]$. $\diamondsuit$

30.3 Variation of the Symplectic Volume

Let (M, ω, S^1, μ) be a hamiltonian S^1-space of dimension $2n$ and let (M_x, ω_x) be its reduced space at level x. Proposition 30.7 or Theorem 30.8 imply that, for x in a sufficiently narrow neighborhood of 0, the symplectic volume of M_x,

$$\text{vol}(M_x) = \int_{M_x} \frac{\omega_x^{n-1}}{(n-1)!} = \int_{M_{\text{red}}} \frac{(\omega_{\text{red}} - x\beta)^{n-1}}{(n-1)!} \ ,$$

is a polynomial in x of degree $n - 1$. This volume can be also expressed as

$$\text{vol}(M_x) = \int_Z \frac{\pi^*(\omega_{\text{red}} - x\beta)^{n-1}}{(n-1)!} \wedge \alpha .$$

Recall that α is a chosen connection form for the S^1-bundle $Z \to M_{\text{red}}$ and β is its curvature form.

Now we go back to the computation of the Duistermaat-Heckman measure. For a Borel subset U of $(-\varepsilon, \varepsilon)$, the Duistermaat-Heckman measure is, by definition,

$$m_{DH}(U) = \int_{\mu^{-1}(U)} \frac{\omega^n}{n!} .$$

Using the fact that $(\mu^{-1}(-\varepsilon, \varepsilon), \omega)$ is symplectomorphic to $(Z \times (-\varepsilon, \varepsilon), \sigma)$ and, moreover, they are isomorphic as hamiltonian S^1-spaces, we obtain

$$m_{DH}(U) = \int_{Z \times U} \frac{\sigma^n}{n!} .$$

Since $\sigma = \pi^* \omega_{\text{red}} - d(x\alpha)$, its power is

$$\sigma^n = n(\pi^* \omega_{\text{red}} - x d\alpha)^{n-1} \wedge \alpha \wedge dx .$$

By the Fubini theorem, we then have

$$m_{DH}(U) = \int_U \left[\int_Z \frac{\pi^*(\omega_{\text{red}} - x\beta)^{n-1}}{(n-1)!} \wedge \alpha \right] \wedge dx .$$

Therefore, the Radon-Nikodym derivative of m_{DH} with respect to the Lebesgue measure, dx, is

$$f(x) = \int_Z \frac{\pi^*(\omega_{\text{red}} - x\beta)^{n-1}}{(n-1)!} \wedge \alpha = \text{vol}(M_x) .$$

The previous discussion proves that, for $x \approx 0$, $f(x)$ is a polynomial in x. The same holds for a neighborhood of any other regular value of μ, because we may change the moment map μ by an arbitrary additive constant.

Homework 23: S^1-Equivariant Cohomology

1. Let M be a manifold with a circle action and $X^\#$ the vector field on M generated by S^1. The algebra of S^1-**equivariant forms on** M is the algebra of S^1-invariant forms on M tensored with complex polynomials in x,

$$\Omega^\bullet_{S^1}(M) := (\Omega^\bullet(M))^{S^1} \otimes_\mathbb{R} \mathbb{C}[x] \ .$$

The product $\wedge$ on $\Omega^\bullet_{S^1}(M)$ combines the wedge product on $\Omega^\bullet(M)$ with the product of polynomials on $\mathbb{C}[x]$.

(a) We grade $\Omega^\bullet_{S^1}(M)$ by adding the usual grading on $\Omega^\bullet(M)$ to a grading on $\mathbb{C}[x]$ where the monomial x has degree 2. Check that $(\Omega^\bullet_{S^1}(M), \wedge)$ is then a supercommutative graded algebra, i.e.,

$$\underline{\alpha} \wedge \underline{\beta} = (-1)^{\deg \underline{\alpha} \cdot \deg \underline{\beta}} \underline{\beta} \wedge \underline{\alpha}$$

for elements of pure degree $\underline{\alpha}, \underline{\beta} \in \Omega^\bullet_{S^1}(M)$.

(b) On $\Omega^\bullet_{S^1}(M)$ we define an operator

$$d_{S^1} := d \otimes 1 - \iota_{X^\#} \otimes x \ .$$

In other words, for an elementary form $\underline{\alpha} = \alpha \otimes p(x)$,

$$d_{S^1} \underline{\alpha} = d\alpha \otimes p(x) - \iota_{X^\#} \alpha \otimes x p(x) \ .$$

The operator d_{S^1} is called the **Cartan differentiation**. Show that d_{S^1} is a superderivation of degree 1, i.e., check that it increases degree by 1 and that it satisfies the *super* Leibniz rule:

$$d_{S^1}(\underline{\alpha} \wedge \underline{\beta}) = (d_{S^1} \underline{\alpha}) \wedge \underline{\beta} + (-1)^{\deg \underline{\alpha}} \underline{\alpha} \wedge d_{S^1} \underline{\beta} \ .$$

(c) Show that $d^2_{S^1} = 0$.

   Hint: Cartan magic formula.

2. The previous exercise shows that the sequence

$$0 \longrightarrow \Omega^0_{S^1}(M) \xrightarrow{d_{S^1}} \Omega^1_{S^1}(M) \xrightarrow{d_{S^1}} \Omega^2_{S^1}(M) \xrightarrow{d_{S^1}} \dots$$

forms a graded complex whose cohomology is called the **equivariant cohomology**[4] of M for the given action of S^1. The kth equivariant cohomology

---

[4] The **equivariant cohomology** of a topological space M endowed with a continuous action of a topological group G is, by definition, the cohomology of the diagonal quotient $(M \times EG)/G$, where EG is the *universal bundle* of G, i.e., EG is a contractible space where G acts freely. H. Cartan [21, 59] showed that, for the action of a compact Lie group G on a manifold M, the de Rham model $(\Omega^\bullet_G(M), d_G)$ computes the equivariant cohomology, where $\Omega^\bullet_G(M)$ are the G-equivariant

group of M is

$$H^k_{S^1}(M) := \frac{\ker d_{S^1} : \Omega^k_{S^1} \longrightarrow \Omega^{k+1}_{S^1}}{\operatorname{im} d_{S^1} : \Omega^{k-1}_{S^1} \longrightarrow \Omega^k_{S^1}} \ .$$

(a) What is the equivariant cohomology of a point?
(b) What is the equivariant cohomology of S^1 with its multiplication action on itself?
(c) Show that the equivariant cohomology of a manifold M with a free S^1-action is isomorphic to the ordinary cohomology of the quotient space M/S^1.

> **Hint:** Let $\pi : M \to M/S^1$ be projection. Show that
>
> $$\pi^* : H^\bullet(M/S^1) \longrightarrow H^\bullet_{S^1}(M)$$
> $$[\alpha] \longmapsto [\pi^*\alpha \otimes 1]$$
>
> is a well-defined isomorphism. It helps to choose a connection on the principal S^1-bundle $M \to M/S^1$, that is, a 1-form θ on M such that $\mathcal{L}_{X^\#}\theta = 0$ and $\iota_{X^\#}\theta = 1$. Keep in mind that a form β on M is of type $\pi^*\alpha$ for some α if and only if it is *basic*, that is $\mathcal{L}_{X^\#}\beta = 0$ and $\iota_{X^\#}\beta = 0$.

3. Suppose that (M, ω) is a symplectic manifold with an S^1-action. Let $\mu \in C^\infty(M)$ be a real function. Consider the equivariant form

$$\underline{\omega} := \omega \otimes 1 + \mu \otimes x \ .$$

Show that $\underline{\omega}$ is **equivariantly closed**, i.e., $d_{S^1}\underline{\omega} = 0$ if and only if μ is a moment map. The equivariant form $\underline{\omega}$ is called the **equivariant symplectic form**.

4. Let M^{2n} be a compact oriented manifold, not necessarily symplectic, acted upon by S^1. Suppose that the set M^{S^1} of fixed points for this action is finite. Let $\alpha^{(2n)}$ be an S^1-invariant form which is the top degree part of an equivariantly closed form of even degree, that is, $\alpha^{(2n)} \in \Omega^{2n}(M)^{S^1}$ is such that there exists $\underline{\alpha} \in \Omega^\bullet_{S^1}(M)$ with

$$\underline{\alpha} = \alpha^{(2n)} + \alpha^{(2n-2)} + \ldots + \alpha^{(0)}$$

where $\alpha^{(2k)} \in (\Omega^{2k}(M))^{S^1} \otimes \mathbb{C}[x]$ and $d_{S^1}\underline{\alpha} = 0$.

a. Show that the restriction of $\alpha^{(2n)}$ to $M \setminus M^{S^1}$ is exact.

> **Hint:** The generator $X^\#$ of the S^1-action does not vanish on $M \setminus M^{S^1}$. Hence, we can define a *connection* on $M \setminus M^{S^1}$ by $\theta(Y) = \frac{\langle Y, X^\# \rangle}{\langle X^\#, X^\# \rangle}$, where $\langle \cdot, \cdot \rangle$ is some S^1-invariant metric on M. Use $\theta \in \Omega^1(M \setminus M^{S^1})$ to chase the primitive of $\alpha^{(2n)}$ all the way up from $\alpha^{(0)}$.

b. Compute the integral of $\alpha^{(2n)}$ over M.

> **Hint:** Stokes' theorem allows to *localize* the answer near the fixed points.

---

forms on M. [8, 9, 29, 54] explain equivariant cohomology in the symplectic context and [59] discusses equivariant de Rham theory and many applications.

This exercise is a very special case of the Atiyah-Bott-Berline-Vergne localization theorem for equivariant cohomology [8, 14].

5. What is the integral of the symplectic form ω on a surface with a hamiltonian S^1-action, knowing that the S^1-action is free outside a finite set of fixed points?

   Hint: Exercises 3 and 4.

References

1. Abraham, R., Marsden, J. E., *Foundations of Mechanics*, second edition, Addison-Wesley, Reading, 1978.
2. Aebischer, B., Borer, M., Kälin, M., Leuenberger, Ch., Reimann, H.M., *Symplectic Geometry. An Introduction Based on the Seminar in Bern, 1992*, Progress in Mathematics **124**, Birkhäuser Verlag, Basel, 1994.
3. Arnold, V., *Mathematical Methods of Classical Mechanics*, Graduate Texts in Math. **60**, Springer-Verlag, New York, 1978.
4. Arnold, V., First steps of symplectic topology, *VIIIth International Congress on Mathematical Physics (Marseille, 1986)*, 1–16, World Sci. Publishing, Singapore, 1987.
5. Arnold, V., Givental, A., Symplectic geometry, *Dynamical Systems IV, Symplectic Geometry and its Applications*, edited by Arnold, V. and Novikov, S., Encyclopaedia of Mathematical Sciences **4**, Springer-Verlag, Berlin-New York, 1990.
6. Atiyah, M., Convexity and commuting Hamiltonians, *Bull. London Math. Soc.* **14** (1982), 1–15.
7. Atiyah, M., Bott, R., The Yang-Mills equations over Riemann surfaces, *Topology* **23** (1984), 1–28. *Philos. Trans. Roy. Soc. London* **308** (1983), 523–615.
8. Atiyah, M., Bott, R., The moment map and equivariant cohomology, *Topology* **23** (1984), 1–28.
9. Audin, M., *The Topology of Torus Actions on Symplectic Manifolds*, Progress in Mathematics **93**, Birkhäuser Verlag, Basel, 1991.
10. Audin, M., *Spinning Tops. A Course on Integrable Systems*, Cambridge Studies in Advanced Mathematics **51**, Cambridge University Press, Cambridge, 1996.
11. Audin, M., Lafontaine, J., Eds., *Holomorphic Curves in Symplectic Geometry*, Progress in Mathematics **117**, Birkhäuser Verlag, Basel, 1994.
12. Auroux, D., Asymptotically holomorphic families of symplectic submanifolds, *Geom. Funct. Anal.* **7** (1997), 971–995.
13. Berline, N., Getzler, E., Vergne, M., *Heat Kernels and Dirac Operators*, Grundlehren der Mathematischen Wissenschaften **298**, Springer-Verlag, Berlin, 1992.
14. Berline, N., Vergne, M., Classes caractéristiques équivariantes, formule de localisation en cohomologie équivariante, *C. R. Acad. Sci. Paris Sér. I Math.* **295** (1982), 539–541.
15. Berline, N., Vergne, M., Zéros d'un champ de vecteurs et classes caractéristiques équivariantes, *Duke Math. J.* **50** (1983), 539–549.
16. Biran, P., A stability property of symplectic packing, *Invent. Math.* **136** (1999), 123–155.
17. Bredon, G., *Introduction to Compact Transformation Groups*, Pure and Applied Mathematics **46**, Academic Press, New York-London, 1972.
18. Bott, R., Tu, L., *Differential Forms in Algebraic Topology*, Graduate Texts in Mathematics **82**, Springer-Verlag, New York-Berlin, 1982.

19. Cannas da Silva, A., Guillemin, V., Woodward, C., On the unfolding of folded symplectic structures, *Math. Res. Lett.* **7** (2000), 35–53.
20. Cannas da Silva, A., Weinstein, A., *Geometric Models for Noncommutative Algebras*, Berkeley Mathematics Lecture Notes series, Amer. Math. Soc., Providence, 1999.
21. Cartan, H., La transgression dans un groupe de Lie et dans un espace fibré principal, *Colloque de Topologie (Espaces Fibrés)*, Bruxelles, 1950, 57–71, Masson et Cie., Paris, 1951.
22. Chern, S.S., *Complex Manifolds Without Potential Theory*, with an appendix on the geometry of characteristic classes, second edition, Universitext, Springer-Verlag, New York-Heidelberg, 1979.
23. Delzant, T., Hamiltoniens périodiques et images convexes de l'application moment, *Bull. Soc. Math. France* **116** (1988), 315–339.
24. Donaldson, S., Symplectic submanifolds and almost-complex geometry, *J. Differential Geom.* **44** (1996), 666–705.
25. Donaldson, S., Lefschetz fibrations in symplectic geometry, Proceedings of the International Congress of Mathematicians, vol. II (Berlin, 1998), *Doc. Math.* **1998**, extra vol. II, 309–314.
26. Donaldson, S., Lefschetz pencils on symplectic manifolds, *J. Differential Geom.* **53** (1999), 205–236.
27. Donaldson, S., Kronheimer, P., *The Geometry of Four-Manifolds*, Oxford Mathematical Monographs, The Clarendon Press, Oxford University Press, New York, 1990.
28. Duistermaat, J.J., On global action-angle coordinates, *Comm. Pure Appl. Math.* **33** (1980), 687–706.
29. Duistermaat, J.J., Equivariant cohomology and stationary phase, *Symplectic Geometry and Quantization* (Sanda and Yokohama, 1993), edited by Maeda, Y., Omori, H. and Weinstein, A., 45-62, *Contemp. Math.* **179**, Amer. Math. Soc., Providence, 1994.
30. Duistermaat, J.J., *The Heat Kernel Lefschetz Fixed Point Formula for the Spin-c Dirac Operator*, Progress in Nonlinear Differential Equations and their Applications **18**, Birkhäuser Boston, Inc., Boston, 1996.
31. Duistermaat, J.J., Heckman, G., On the variation in the cohomology of the symplectic form of the reduced phase space, *Invent. Math.* **69** (1982), 259–268; Addendum, *Invent. Math.* **72** (1983), 153–158.
32. Eliashberg, Y., Classification of overtwisted contact structures on 3-manifolds, *Invent. Math.* **98** (1989), 623–637.
33. Eliashberg, Y., Contact 3-manifolds twenty years since J. Martinet's work, *Ann. Inst. Fourier (Grenoble)* **42** (1992), 165–192.
34. Eliashberg, Y., Gromov, M., Lagrangian intersection theory: finite-dimensional approach, *Geometry of Differential Equations*, 27–118, *Amer. Math. Soc. Transl. Ser. 2*, **186**, Amer. Math. Soc., Providence, 1998.
35. Eliashberg, Y., Thurston, W., *Confoliations*, University Lecture Series **13**, Amer. Math. Soc., Providence, 1998.
36. Eliashberg, Y., Traynor, L., Eds., *Symplectic Geometry and Topology*, lectures from the Graduate Summer School held in Park City, June 29-July 19, 1997, IAS/Park City Mathematics Series **7**, Amer. Math. Soc., Providence, 1999.
37. Fernández, M., Gotay, M., Gray, A., Compact parallelizable four-dimensional symplectic and complex manifolds, *Proc. Amer. Math. Soc.* **103** (1988), 1209–1212.
38. Fulton, W., *Introduction to Toric Varieties*, Annals of Mathematics Studies **131**, Princeton University Press, Princeton, 1993.
39. Geiges, H., Applications of contact surgery, *Topology* **36** (1997), 1193–1220.
40. Geiges, H., Gonzalo, J., Contact geometry and complex surfaces, *Invent. Math.* **121** (1995), 147–209.
41. Ginzburg, V., Guillemin, V., Karshon, Y., Cobordism theory and localization formulas for Hamiltonian group actions, *Internat. Math. Res. Notices* 1996, 221–234.
42. Ginzburg, V., Guillemin, V., Karshon, Y., The relation between compact and non-compact equivariant cobordisms, *Tel Aviv Topology Conference: Rothenberg Festschrift (1998)*, 99–112, Contemp. Math., **231**, Amer. Math. Soc., Providence, 1999.

43. Giroux, E., Topologie de contact en dimension 3 (autour des travaux de Yakov Eliashberg), *Séminaire Bourbaki* **1992/93**, *Astérisque* **216** (1993), 7–33.
44. Giroux, E., Une structure de contact, même tendue, est plus ou moins tordue, *Ann. Sci. École Norm. Sup.* (4) **27** (1994), 697–705.
45. Givental, A., Periodic mappings in symplectic topology (Russian), *Funktsional. Anal. i Prilozhen* **23** (1989), 37–52, translation in *Funct. Anal. Appl.* **23** (1989), 287–300.
46. Gompf, R., A new construction of symplectic manifolds, *Ann. of Math.* **142** (1995), 527–595.
47. Gotay, M., On coisotropic imbeddings of presymplectic manifolds, *Proc. Amer. Math. Soc.* **84** (1982), 111–114.
48. Griffiths, P., Harris, J., *Principles of Algebraic Geometry*, Chapter 0, reprint of the 1978 original, Wiley Classics Library, John Wiley & Sons, Inc., New York, 1994.
49. Gromov, M., Pseudoholomorphic curves in symplectic manifolds, *Invent. Math.* **82** (1985), 307–347.
50. Gromov, M., *Partial Differential Relations*, Springer-Verlag, Berlin-New York, 1986.
51. Gromov, M., Soft and hard symplectic geometry, *Proceedings of the International Congress of Mathematicians* **1** (Berkeley, Calif., 1986), 81-98, Amer. Math. Soc., Providence, 1987.
52. Guggenheimer, H., Sur les variétés qui possèdent une forme extérieure quadratique fermée, *C. R. Acad. Sci. Paris* **232** (1951), 470–472.
53. Guillemin, V., Course 18.966 – Geometry of Manifolds, M.I.T., Spring of 1992.
54. Guillemin, V., *Moment Maps and Combinatorial Invariants of Hamiltonian T^n-spaces*, Progress in Mathematics **122**, Birkhäuser, Boston, 1994.
55. Guillemin, V., Pollack, A., *Differential Topology*, Prentice-Hall, Inc., Englewood Cliffs, N.J., 1974.
56. Guillemin, V., Sternberg, S., *Geometric Asymptotics*, Math. Surveys and Monographs **14**, Amer. Math. Soc., Providence, 1977.
57. Guillemin, V., Sternberg, S., Convexity properties of the moment mapping, *Invent. Math.* **67** (1982), 491–513.
58. Guillemin, V., Sternberg, S., *Symplectic Techniques in Physics*, second edition, Cambridge University Press, Cambridge, 1990.
59. Guillemin, V., Sternberg, S., *Supersymmetry and Equivariant de Rham Theory*, with an appendix containing two reprints by H. Cartan, Mathematics Past and Present, Springer-Verlag, Berlin, 1999.
60. Hausmann, J.-C., Knutson, A., The cohomology ring of polygon spaces, *Ann. Inst. Fourier (Grenoble)* **48** (1998), 281–321.
61. Hausmann, J.-C., Knutson, A., Cohomology rings of symplectic cuts, *Differential Geom. Appl.* **11** (1999), 197–203.
62. Hitchin, N., Segal, G., Ward, R., *Integrable Systems. Twistors, Loop groups, and Riemann Surfaces* Oxford Graduate Texts in Mathematics **4**, The Clarendon Press, Oxford University Press, New York, 1999.
63. Hofer, H., Pseudoholomorphic curves in symplectizations with applications to the Weinstein conjecture in dimension three, *Invent. Math.* **114** (1993), 515–563.
64. Hofer, H. Viterbo, C., The Weinstein conjecture in the presence of holomorphic spheres, *Comm. Pure Appl. Math.* **45** (1992), 583–622.
65. Hofer, H., Zehnder, E., *Symplectic Invariants and Hamiltonian Dynamics*, Birkhäuser Advanced Texts: Basler Lehrbücher, Birkhäuser Verlag, Basel, 1994.
66. Hörmander, L., *An Introduction to Complex Analysis in Several Variables*, third edition, North-Holland Mathematical Library **7**, North-Holland Publishing Co., Amsterdam-New York, 1990.
67. Jacobson, N., *Lie Algebras*, republication of the 1962 original, Dover Publications, Inc., New York, 1979.
68. Jeffrey, L., Kirwan, F., Localization for nonabelian group actions, *Topology* **34** (1995), 291–327.
69. Kirwan, F., *Cohomology of Quotients in Symplectic and Algebraic Geometry*, Mathematical Notes **31**, Princeton University Press, Princeton, 1984.

70. Kodaira, K., On the structure of compact complex analytic surfaces, I, *Amer. J. Math.* **86** (1964), 751–798.

71. Kronheimer, P., Developments in symplectic topology, *Current Developments in Mathematics* (Cambridge, 1998), 83–104, Int. Press, Somerville, 1999.

72. Kronheimer, P., Mrowka, T., Monopoles and contact structures, *Invent. Math.* **130** (1997), 209–255.

73. Lalonde, F., McDuff, D., The geometry of symplectic energy, *Ann. of Math.* (2) **141** (1995), 349–371.

74. Lalonde, F., Polterovich, L., Symplectic diffeomorphisms as isometries of Hofer's norm, *Topology* **36** (1997), 711–727.

75. Lerman, E., Meinrenken, E., Tolman, S., Woodward, C., Nonabelian convexity by symplectic cuts, *Topology* **37** (1998), 245–259.

76. Marsden, J., Ratiu, T., *Introduction to Mechanics and Symmetry. A Basic Exposition of Classical Mechanical Systems*, Texts in Applied Mathematics **17**, Springer-Verlag, New York, 1994.

77. Marsden, J., Weinstein, A., Reduction of symplectic manifolds with symmetry, *Rep. Mathematical Phys.* **5** (1974), 121–130.

78. Martin, S., Symplectic quotients by a nonabelian group and by its maximal torus, arXiv:math/0001002.

79. Martin, S., Transversality theory, cobordisms, and invariants of symplectic quotients, arXiv:math/0001001.

80. Martinet, J., Formes de contact sur les variétés de dimension 3, *Proceedings of Liverpool Singularities Symposium* II (1969/1970), 142-163, Lecture Notes in Math. **209**, Springer, Berlin, 1971.

81. McDuff, D., Examples of simply-connected symplectic non-Kählerian manifolds, *J. Differential Geom.* **20** (1984), 267–277.

82. McDuff, D., The local behaviour of holomorphic curves in almost complex 4-manifolds, *J. Differential Geom.* **34** (1991), 143–164.

83. McDuff, D., Salamon, D., *Introduction to Symplectic Topology*, Oxford Mathematical Monographs, Oxford University Press, New York, 1995.

84. Meinrenken, E., Woodward, C., Hamiltonian loop group actions and Verlinde factorization, *J. Differential Geom.* **50** (1998), 417–469.

85. Meyer, K., Symmetries and integrals in mechanics, *Dynamical Systems* (Proc. Sympos., Univ. Bahia, Salvador, 1971), 259–272. Academic Press, New York, 1973.

86. Milnor, J., *Morse Theory*, based on lecture notes by M. Spivak and R. Wells, Annals of Mathematics Studies **51**, Princeton University Press, Princeton, 1963.

87. Moser, J., On the volume elements on a manifold, *Trans. Amer. Math. Soc.* **120** (1965), 286-294.

88. Mumford, D., Fogarty, J., Kirwan, F., *Geometric Invariant Theory*, Ergebnisse der Mathematik und ihrer Grenzgebiete **34**, Springer-Verlag, Berlin, 1994.

89. Newlander, A., Nirenberg, L., Complex analytic coordinates in almost complex manifolds, *Ann. of Math.* **65** (1957), 391–404.

90. Salamon, D., Morse theory, the Conley index and Floer homology, *Bull. London Math. Soc.* **22** (1990), 113–140.

91. Satake, I., On a generalization of the notion of manifold, *Proc. Nat. Acad. Sci. U.S.A.* **42** (1956), 359–363.

92. Scott, P., The geometries of 3-manifolds, *Bull. London Math. Soc.* **15** (1983), 401–487.

93. Seidel, P., Lagrangian two-spheres can be symplectically knotted, *J. Differential Geom.* **52** (1999), 145–171.

94. Sjamaar, R., Lerman, E., Stratified symplectic spaces and reduction, *Ann. of Math.* **134** (1991), 375–422.

95. Souriau, J.-M. , *Structure des Systèmes Dynamiques*, Maîtrises de Mathématiques, Dunod, Paris 1970.

96. Spivak, M., *A Comprehensive Introduction to Differential Geometry*, Vol. I, second edition, Publish or Perish, Inc., Wilmington, 1979.

97. Taubes, C., The Seiberg-Witten invariants and symplectic forms, *Math. Res. Lett.* **1** (1994), 809–822.

98. Taubes, C., More constraints on symplectic forms from Seiberg-Witten invariants, *Math. Res. Lett.* **2** (1995), 9–13.

99. Taubes, C., The Seiberg-Witten and Gromov invariants, *Math. Res. Lett.* **2** (1995), 221–238.

100. Thomas, C., Eliashberg, Y., Giroux, E., 3-dimensional contact geometry, *Contact and Symplectic Geometry (Cambridge, 1994)*, 48-65, Publ. Newton Inst. **8**, Cambridge University Press, Cambridge, 1996.

101. Thurston, W., Some simple examples of symplectic manifolds, *Proc. Amer. Math. Soc.* **55** (1976), 467–468.

102. Tolman, S., Weitsman, J., On semifree symplectic circle actions with isolated fixed points, *Topology* **39** (2000), 299–309.

103. Viterbo, C., A proof of Weinstein's conjecture in $\mathbb{R}^{2n}$, *Ann. Inst. H. Poincaré Anal. Non Linéaire* **4** (1987), 337–356.

104. Weinstein, A., Symplectic manifolds and their Lagrangian submanifolds, *Advances in Math.* **6** (1971), 329–346.

105. Weinstein, A., *Lectures on Symplectic Manifolds*, Regional Conference Series in Mathematics **29**, Amer. Math. Soc., Providence, 1977.

106. Weinstein, A., On the hypotheses of Rabinowitz' periodic orbit theorems, *J. Differential Equations* **33** (1979), 353–358.

107. Weinstein, A., Neighborhood classification of isotropic embeddings, *J. Differential Geom.* **16** (1981), 125–128.

108. Weinstein, A., Symplectic geometry, *Bull. Amer. Math. Soc. (N.S.)* **5** (1981), 1–13.

109. Wells, R.O., *Differential Analysis on Complex Manifolds*, second edition, Graduate Texts in Mathematics **65**, Springer-Verlag, New York-Berlin, 1980.

110. Weyl, H., *The Classical Groups. Their Invariants and Representations*, Princeton Landmarks in Mathematics, Princeton University Press, Princeton, 1997.

111. Witten, E., Two-dimensional gauge theories revisited, *J. Geom. Phys.* **9** (1992), 303–368.

112. Woodward, C., Multiplicity-free Hamiltonian actions need not be Kähler, *Invent. Math.* **131** (1998), 311–319.

113. Woodward, C., Spherical varieties and existence of invariant Kähler structures, *Duke Math. J.* **93** (1998), 345–377.

Index

Lecture Notes in Mathematics

For information about earlier volumes
please contact your bookseller or Springer
LNM Online archive: springerlink.com

Vol. 1855: G. Da Prato, P.C. Kunstmann, I. Lasiecka, A. Lunardi, R. Schnaubelt, L. Weis, Functional Analytic Methods for Evolution Equations. Editors: M. Iannelli, R. Nagel, S. Piazzera (2004)

Vol. 1856: K. Back, T.R. Bielecki, C. Hipp, S. Peng, W. Schachermayer, Stochastic Methods in Finance, Bressanone/Brixen, Italy, 2003. Editors: M. Fritelli, W. Runggaldier (2004)

Vol. 1857: M. Émery, M. Ledoux, M. Yor (Eds.), Séminaire de Probabilités XXXVIII (2005)

Vol. 1858: A.S. Cherny, H.-J. Engelbert, Singular Stochastic Differential Equations (2005)

Vol. 1859: E. Letellier, Fourier Transforms of Invariant Functions on Finite Reductive Lie Algebras (2005)

Vol. 1860: A. Borisyuk, G.B. Ermentrout, A. Friedman, D. Terman, Tutorials in Mathematical Biosciences I. Mathematical Neurosciences (2005)

Vol. 1861: G. Benettin, J. Henrard, S. Kuksin, Hamiltonian Dynamics – Theory and Applications, Cetraro, Italy, 1999. Editor: A. Giorgilli (2005)

Vol. 1862: B. Helffer, F. Nier, Hypoelliptic Estimates and Spectral Theory for Fokker-Planck Operators and Witten Laplacians (2005)

Vol. 1863: H. Führ, Abstract Harmonic Analysis of Continuous Wavelet Transforms (2005)

Vol. 1864: K. Efstathiou, Metamorphoses of Hamiltonian Systems with Symmetries (2005)

Vol. 1865: D. Applebaum, B.V. R. Bhat, J. Kustermans, J. M. Lindsay, Quantum Independent Increment Processes I. From Classical Probability to Quantum Stochastic Calculus. Editors: M. Schürmann, U. Franz (2005)

Vol. 1866: O.E. Barndorff-Nielsen, U. Franz, R. Gohm, B. Kümmerer, S. Thorbjønsen, Quantum Independent Increment Processes II. Structure of Quantum Lévy Processes, Classical Probability, and Physics. Editors: M. Schürmann, U. Franz, (2005)

Vol. 1867: J. Sneyd (Ed.), Tutorials in Mathematical Biosciences II. Mathematical Modeling of Calcium Dynamics and Signal Transduction. (2005)

Vol. 1868: J. Jorgenson, S. Lang, $Pos_n(R)$ and Eisenstein Series. (2005)

Vol. 1869: A. Dembo, T. Funaki, Lectures on Probability Theory and Statistics. Ecole d'Eté de Probabilités de Saint-Flour XXXIII-2003. Editor: J. Picard (2005)

Vol. 1870: V.I. Gurariy, W. Lusky, Geometry of Müntz Spaces and Related Questions. (2005)

Vol. 1871: P. Constantin, G. Gallavotti, A.V. Kazhikhov, Y. Meyer, S. Ukai, Mathematical Foundation of Turbulent Viscous Flows, Martina Franca, Italy, 2003. Editors: M. Cannone, T. Miyakawa (2006)

Vol. 1872: A. Friedman (Ed.), Tutorials in Mathematical Biosciences III. Cell Cycle, Proliferation, and Cancer (2006)

Vol. 1873: R. Mansuy, M. Yor, Random Times and Enlargements of Filtrations in a Brownian Setting (2006)

Vol. 1874: M. Yor, M. Émery (Eds.), In Memoriam Paul-André Meyer - Séminaire de Probabilités XXXIX (2006)

Vol. 1875: J. Pitman, Combinatorial Stochastic Processes. Ecole d'Eté de Probabilités de Saint-Flour XXXII-2002. Editor: J. Picard (2006)

Vol. 1876: H. Herrlich, Axiom of Choice (2006)

Vol. 1877: J. Steuding, Value Distributions of L-Functions (2007)

Vol. 1878: R. Cerf, The Wulff Crystal in Ising and Percolation Models, Ecole d'Eté de Probabilités de Saint-Flour XXXIV-2004. Editor: Jean Picard (2006)

Vol. 1879: G. Slade, The Lace Expansion and its Applications, Ecole d'Eté de Probabilités de Saint-Flour XXXIV-2004. Editor: Jean Picard (2006)

Vol. 1880: S. Attal, A. Joye, C.-A. Pillet, Open Quantum Systems I, The Hamiltonian Approach (2006)

Vol. 1881: S. Attal, A. Joye, C.-A. Pillet, Open Quantum Systems II, The Markovian Approach (2006)

Vol. 1882: S. Attal, A. Joye, C.-A. Pillet, Open Quantum Systems III, Recent Developments (2006)

Vol. 1883: W. Van Assche, F. Marcellàn (Eds.), Orthogonal Polynomials and Special Functions, Computation and Application (2006)

Vol. 1884: N. Hayashi, E.I. Kaikina, P.I. Naumkin, I.A. Shishmarev, Asymptotics for Dissipative Nonlinear Equations (2006)

Vol. 1885: A. Telcs, The Art of Random Walks (2006)

Vol. 1886: S. Takamura, Splitting Deformations of Degenerations of Complex Curves (2006)

Vol. 1887: K. Habermann, L. Habermann, Introduction to Symplectic Dirac Operators (2006)

Vol. 1888: J. van der Hoeven, Transseries and Real Differential Algebra (2006)

Vol. 1889: G. Osipenko, Dynamical Systems, Graphs, and Algorithms (2006)

Vol. 1890: M. Bunge, J. Funk, Singular Coverings of Toposes (2006)

Vol. 1891: J.B. Friedlander, D.R. Heath-Brown, H. Iwaniec, J. Kaczorowski, Analytic Number Theory, Cetraro, Italy, 2002. Editors: A. Perelli, C. Viola (2006)

Vol. 1892: A. Baddeley, I. Bárány, R. Schneider, W. Weil, Stochastic Geometry, Martina Franca, Italy, 2004. Editor: W. Weil (2007)

Vol. 1893: H. Hanßmann, Local and Semi-Local Bifurcations in Hamiltonian Dynamical Systems, Results and Examples (2007)

Vol. 1894: C.W. Groetsch, Stable Approximate Evaluation of Unbounded Operators (2007)

Vol. 1895: L. Molnár, Selected Preserver Problems on Algebraic Structures of Linear Operators and on Function Spaces (2007)

Vol. 1896: P. Massart, Concentration Inequalities and Model Selection, Ecole d'Été de Probabilités de Saint-Flour XXXIII-2003. Editor: J. Picard (2007)

Vol. 1897: R. Doney, Fluctuation Theory for Lévy Processes, Ecole d'Été de Probabilités de Saint-Flour XXXV-2005. Editor: J. Picard (2007)

Vol. 1898: H.R. Beyer, Beyond Partial Differential Equations, On linear and Quasi-Linear Abstract Hyperbolic Evolution Equations (2007)

Vol. 1899: Séminaire de Probabilités XL. Editors: C. Donati-Martin, M. Émery, A. Rouault, C. Stricker (2007)

Vol. 1900: E. Bolthausen, A. Bovier (Eds.), Spin Glasses (2007)

Vol. 1901: O. Wittenberg, Intersections de deux quadriques et pinceaux de courbes de genre 1, Intersections of Two Quadrics and Pencils of Curves of Genus 1 (2007)

Vol. 1902: A. Isaev, Lectures on the Automorphism Groups of Kobayashi-Hyperbolic Manifolds (2007)

Vol. 1903: G. Kresin, V. Maz'ya, Sharp Real-Part Theorems (2007)

Vol. 1904: P. Giesl, Construction of Global Lyapunov Functions Using Radial Basis Functions (2007)

Vol. 1905: C. Prévôt, M. Röckner, A Concise Course on Stochastic Partial Differential Equations (2007)

Recent Reprints and New Editions